Zentrum für
Umweltforschung und
Umwelttechnologie

Worms and wetland water:
The role of lumbricids and enchytraeids
in nutrient release from flooded grassland ecosystems

Dissertation zur Erlangung des naturwissenschaftlichen Doktorgrades
im Fach Biologie an der Universität Bremen

vorgelegt von

Nathalie Madeleine Plum
Eutzinger Str. 34
76829 Landau i.d. Pfalz

1. Gutachterin: Prof. Dr. Juliane Filser
2. Gutachter: Dr. Michael Schirmer

Bibliografische Information Der Deutschen Bibliothek

Die Deutsche Bibliothek verzeichnet diese Publikation in der Deutschen
Nationalbibliografie; detaillierte bibliografische Daten sind im Internet über
http://dnb.ddb.de abrufbar.

ISBN 3-8325-1105-9

Logos Verlag Berlin
Comeniushof, Gubener Str. 47,
10243 Berlin
Tel.: +49 030 42 85 10 90
Fax: +49 030 42 85 10 92
INTERNET: http://www.logos-verlag.de

Table of Contents

Abbreviations and Symbols

Abbreviation	Meaning
A. caliginosa	*Aporrectodea caliginosa*
A. chlorotica	*Allolobophora chlorotica*
AITC	allyl isothiocyanate (Allylisothiocyanat)
ANOVA	analysis of variance (Varianzanalyse)
C	carbon (Kohlenstoff)
cf.	compare (*conferre;* siehe)
C/N	relation of carbon to nitrogen in soil (Verhältnis von Kohlenstoff zu Stickstoff im Boden)
Ca^{2+}	calcium ions (-Ionen)
$CaCl_2$	Calcium chloride (Calciumchlorid)
C_{mic}	microbial carbon (mikrobiell gebundener Kohlenstoff)
C_{org}	organic carbon content in soil (organ. Kohlenstoffgehalt im Boden)
df	degrees of freedom (Freiheitsgrade)
DON	dissolved organic carbon (gelöster organischer Kohlenstoff)
dw	dry weight (Trockengewicht)
F	F value (Wert) in ANOVA
g	gram (Gramm)
$g\ l^{-1}$	gram per liter (Gramm pro Liter)
$g\ m^{-2}$	gram per square meter (Gramm pro Quadratmeter)
h	hour (Stunde)
H	H value (Wert) in Kruskal-Wallis-Test
ISO	International Standardization Organization (Internationale Standardisierungs-Organisation)
K	potassium (Kalium)
$kg\ ha^{-1}$	kilogram per hectare (Kilogramm pro Hektar)
NLfB	Niedersächsisches Landesamt für Bodenforschung
$ind.\ m^{-2}$	individuals per square meter (Individuen pro Quadratmeter)
juv.	Juvenile (juvenil)
L. rubellus	*Lumbricus rubellus*
L. terrestris	*Lumbricus terrestris*
$mg\ kg^{-1}$	milligram per kilo (Milligramm pro Kilogramm)
Mg^{2+}	magnesium ions (-Ionen)
min.	minutes (Minuten)
mm Hg	millimeter mercury column (Millimeter Quecksilbersäule)
mths	months
$\mu S\ cm^{-1}$	micro Siemens per centimeter (Mikrosiemens pro Zentimeter)
n	number of cases (Anzahl der Fälle)
N	nitrogen (Stickstoff)
N_t	nitrogen (Stickstoff im Boden)
N_2	atmospheric nitrogen (atmosphärischer Stickstoff)
$N\ m^{-2}$	Newton per square meter (pro Quadratmeter)
n.d.	no data (keine Daten)
NH_4^+-N	ammonium-nitrogen (Ammonium-Stickstoff)
N_{min}	mineral nitrogen (mineralischer Stickstoff)
NO_2-N	nitrite-nitrogen (Nitrit-Stickstoff)
NO_3^--N	nitrate-nitrogen (Nitrat-Stickstoff)

NO_x	nitrous oxides (Stickoxide)
ns°	numbers (Nummern)
O. cyaneum	*Octolasion cyaneum*
O. tyrtaeum	*Octolasion tyrtaeum*
p	confidential interval (Vertrauensintervall)
P	phosphorus (Phosphor)
PAH	polycyclic aromatic hydrocarbons (Polyzyklische aromatische Kohlenwasserstoffe)
PE	polyethylene (Polyethylen)
pers. comm.	personal communication (persönliche Mitteilung)
r or R	correlation coefficient (Korrelationskoeffizient)
RM-ANOVA	repeated measurements analysis of variances (Varianzanalyse mit Messwiederholungen)
Rsq	rootsquare (Quadratwurzel)
SD	standard deviation (Standardabweichung)
SE	standard error (Standardfehler)
SOM	soil organic matter
SPSS	Statistical Package for the Social Sciences (Statistik-Programm)
spec.	species, not identified (unbestimmbare Art)
T	T value (Wert) in T-test
vol : fw	volume to fresh weight (Volumen zu Frischgewicht)
vol.-%	volume percent (Volumen-Prozent)
vs.	versus (gegen)
yr	year (Jahr)
Z	Z value (Wert) in Mann-Whitney-U-Test
Ø	diameter (Durchmesser)
% w/ws	maximum water capacity (maximale Wasserhaltekapazität)
% of dw	percent of dry weight (Prozent des Trockengewichts)
°C	degrees (Grad) Celsius

Worms and wetland water: The role of lumbricids and enchytraeids in nutrient release from flooded grassland ecosystems

Summary

The present thesis approaches the question about the role of soil fauna, in particular annelid worms, on nutrient dynamics in three wetlands differing in soil type and flooding dynamics.

General dynamics of nitrogen and phosphorus in wetlands are described. A literature review compares survival strategies and adaptations to flooding of snails and slugs (Gastropoda), earthworms (Lumbricidae), potworms (Enchytraeidae), woodlice (Isopoda), millipedes (Chilopoda, Diplopoda), Diptera, and other insect larvae. Flooding of grassland, especially when frequent or extended over more than three months, reduced diversity, abundance, and biomass of all groups of soil macrofauna.

In a two-year field survey of earthworms and potworms (Oligochaeta: Lumbricidae, Enchytraeidae) in three floodplain meadows in Northern Germany, earthworms dominated the biomass of soil fauna, the dominant species being *Octolasion tyrtaeum* in marsh soil, *Octolasion cyaneum* in peat soil, *Allolobophora chlorotica* in gley soil, and *Lumbricus rubellus* in all three sites. The enchytraeid communities composed of 7-10 species differed widely between sites. While the summer flood of 2002 reduced annelid populations in peat soil to zero, in gley soil only enchytraeids were sharply reduced, whereas earthworms were less affected. The marsh soil was not subjected to summer flooding. Consequently, it had the greatest and most stable annelid populations. During summer drought in 2003, earthworms and enchytraeids were absent from gley soil, while they had maximum abundances in peat soil. The efficiency of the mustard extraction method for earthworms which was used in this field survey depends on site (soil type), earthworm species (life forms), soil humidity, and the state of the earthworms (diapause or active state).

Four laboratory experiments with terrestrial-aquatic microcosms, inoculated with natural densities of earthworms and/or enchytraeids or left as defaunated controls, were conducted to shed light on the annelids' impact on nutrient dynamics in floodwater. During an inundation of 72 hours, the mobilisation of ammonium and total phosphorus was increased in peat soil microcosms with *O. cyaneum* as well as in marsh soil microcosms with *L. rubellus* (both 600 ind. m^{-2}). Enchytraeid treatments

(6000 ind. m^{-2}) showed a trend towards higher nitrate concentrations on peat soil and higher nitrite concentrations on gley soil. Some earthworm effects were only significant in the simultaneous presence of enchytraeids. In a subsequent experiment with peat soil using treatments with different enchytraeid densities up to 16,000 ind. m^{-2} (24 hours of flooding), no enchytraeid effect was significant. However, this experiment confirmed the mobilising effect of *O. cyaneum* on ammonium and phosphorus, even at a comparably low density (200 ind. m^{-2}). Short-term temporal dynamics of nutrient release from peat soil within the first day of flooding showed that the earthworms' effect only becomes visible after 24 hours, even at a high density (800 ind. m^{-2}). The effect of *L. rubellus* (200 ind. m^{-2}) on nutrient mobilisation from peat soil was the same, but less pronounced than that of *O. cyaneum*, while the mobilisation of ammonium and phosphorus from dead tissue of both species was considerable. A subsequent field experiment with mesocosms on the marsh site failed, but showed that in spite of a short flooding time (1.5 hours), ammonium, nitrate, and phosphorus may be mobilised from the soil into the floodwater.

Influences of site, soil type, chemical composition of the floodwater, season, duration of flooding, and experimental design on nutrient dynamics in wetlands are discussed. Effects of annelid worms on nutrient dynamics in wetlands are quantified. Living earthworms (*O. cyaneum* in a maximum density of 8 million ind. ha^{-1}) mobilise up to 0.3 kg ha^{-1} NH$_4^+$-N and 0.16 kg ha^{-1} phosphorus within the first 24 hours of an inundation on peat soil, whereas mobilisation of NH$_4^+$-N and phosphorus from dead tissue (equivalent to 4 million ind. ha^{-1}) reached 0.6 and 2.4 kg ha^{-1}, respectively, while NO$_3^-$-N was reduced by 0.8 kg ha^{-1}. As annelid casts and earthworm mucus are known to be rich in inorganic N and P, the main portion of the mobilised nutrients can be attributed to excrements. Dead earthworm tissue is another important source of ammonium and phosphorus. Further indirect effects of annelids mediated by burrowing, selective grazing, and the fostering of microorganisms involved in nitrification and other processes are discussed.

Provided that the results from the laboratory studies can be validated under field conditions, controlled flooding should guarantee the thriving of earthworm populations with inundations restricted to the winter months and raised water tables in summer. This would provide a solid food base to endangered wading birds and prevent undesired nutrient release into the river water.

Zusammenfassung

Die Rolle von Regenwürmern (Lumbricidae) und Kleinringelwürmern (Enchytraeidae) in der Nährstoffmobilisierung aus Feuchtgrünlandöko-systemen

Die vorliegende Dissertation beschäftigt sich mit der Rolle der Bodenfauna, speziell der Anneliden, in der Nährstoffdynamik von drei Feuchtgrünlandökosystemen, die sich hinsichtlich Bodentyp und Überflutungsdynamik unterscheiden. Die Dynamik von Stickstoff und Phosphor in Feuchtgebieten wird beschrieben. Ein Literatur-Review vergleicht Überlebensstrategien und Anpassungen an Überflutung bei terrestrischen Schnecken (Gastropoda), Regenwürmern (Lumbricidae), Kleinringelwürmern (Enchytraeidae), Asseln (Isopoda), Tausendfüßern (Chilopoda, Diplopoda), Zweiflüglern (Diptera) und anderen Insektenlarven. Die Überflutung von Grünland, insbesondere wenn sie häufig geschieht oder sich auf über drei Monate ausdehnt, reduziert Diversität, Abundanz und Biomasse aller untersuchten Bodentiergruppen. Über einen Zeitraum von zwei Jahren wurde die Populationsdynamik von Regenwürmern und Enchyträen in drei norddeutschen Auen im Freiland untersucht.

Regenwürmer dominierten die Biomasse der Bodenfauna. Die dominanten Arten waren *Octolasion tyrtaeum* im Marschboden, *Octolasion cyaneum* im Torfboden, *Allolobophora chlorotica* im Gleyboden und *Lumbricus rubellus* an allen drei Standorten. Die Enchyträengemeinschaft setzte sich aus jeweils 7 bis 10 Arten zusammen und war je nach Standort sehr unterschiedlich. Während das Sommerhochwasser im Jahr 2002 die Annelidenpopulationen im Torfboden bis auf Null reduzierte, wurden im Gleyboden lediglich die Enchyträen in ihrer Abundanz stark reduziert. Die Regenwürmer hingegen waren sehr viel weniger betroffen. Auf dem Marschboden gab es kein Sommerhochwasser; folglich wurden dort die größten und stabilsten Annelidenpopulationen angetroffen. Während der Sommertrockenheit im Jahr 2003 wurden im Gleyboden keine Anneliden gefunden. Im Torfboden hingegen kamen sie in maximalen Dichten vor. Die Effizienz der Senfextraktion für Regenwürmer, die im Rahmen dieser Freilandarbeiten eingesetzt wurde, hängt ab vom Standort (Bodentyp), von der Regenwurmart (Lebensform), von der Bodenfeuchte sowie vom Zustand der Regenwürmer (Diapause oder aktives Stadium).

Der Einfluss der Anneliden auf die Nährstoffdynamik im Überflutungswasser wurde in vier Laborexperimenten mit terrestrisch-aquatischen Mikrokosmen untersucht. Diese waren mit Regenwürmern und/oder Enchyträen besetzt oder als defaunierte

Kontrollen belassen. Während einer 72-stündigen Überflutung war die Mobilisierung von Ammonium und Gesamtphosphor aus Torfboden erhöht in Anwesenheit von *O. cyaneum*. Die gleiche Beobachtung wurde für den Marschboden in Anwesenheit von *L. rubellus* gemacht (beide Ansätze: 600 ind. m^{-2}). Ansätze mit Enchyträen (6000 ind. m^{-2}) zeigten einen Trend zu höheren Nitratkonzentrationen über Torfboden und höheren Nitritkonzentrationen über Gleyboden. Einige Regenwurmeffekte waren nur signifikant in gleichzeitiger Anwesenheit von Enchyträen. In einem darauf folgenden Experiment mit verschiedenen Enchyträendichten (bis zu 16000 ind. m^{-2}, 24-stündige Überflutung) war kein Enchyträeneffekt signifikant. Dieses Experiment bestätigte jedoch die mobilisierende Wirkung von *O. cyaneum* auf Ammonium und Phosphor, sogar bei einer relativ geringen Dichte (200 ind. m^{-2}). Die Kurzzeitdynamik der Nährstofffreisetzung aus Torfboden innerhalb des ersten Tags einer Überflutung zeigte, dass der Regenwurmeffekt erst nach 24 Stunden erkennbar wird, selbst bei einer hohen Regenwurmdichte (800 ind. m^{-2}). Der Effekt von *L. rubellus* (200 ind. m^{-2}) auf die Nährstoffmobilisierung aus Torfboden war der gleiche, jedoch weniger ausgeprägt als der von *O. cyaneum*, während die Mobilisierung von Ammonium und Phosphor aus totem Regenwurmgewebe beider Arten beachtlich war. Ein anschließendes Freilandexperiment scheiterte, zeigte aber, dass trotz einer kurzen Überflutungszeit (1 ½ Stunden) Ammonium, Nitrat und Phosphor aus dem Marschboden ins Überflutungswasser mobilisiert werden können.

Einflüsse von Standort, Bodentyp, chemischer Zusammensetzung des Überflutungswassers, Jahreszeit, Überflutungsdauer und Experimentdesign auf die Nährstoffdynamik im Feuchtgrünland werden diskutiert. Die Effekte der Anneliden werden quantifiziert. Lebende Regenwürmer (*O. cyaneum* in einer maximalen Dichte von 8 Millionen ind. ha^{-1}) mobilisiert bis zu 0.3 kg ha^{-1} NH$_4^+$-N und 0.16 kg ha^{-1} Phosphor innerhalb der ersten 24 Stunden einer Überflutung auf Torfboden. Die Mobilisierung von NH$_4^+$-N und Phosphor aus totem Gewebe (entsprechend einer Regenwurmdichte von 4 Millionen ind. ha^{-1}) erreichte 0.6 bzw. 2.4 kg ha^{-1}, während NO$_3^-$-N um bis zu 0.8 kg ha^{-1} reduziert wurde. Da Annelidenkot bekanntlich reich an anorganischem N und P ist, kann ein Großteil der mobilisierten Nährstoffe den Exkrementen zugeschrieben werden. Totes Regenwurmgewebe ist ebenfalls eine wichtige Quelle von Ammonium und Phosphor. Weitere (indirekte) Effekte von Anneliden sind denkbar durch ihr Graben, selektiven Fraß und die Förderung von Mikroorganismen, die bei der Nitrifikation und anderen Prozessen beteiligt sind.

Falls die Ergebnisse der Laboruntersuchungen auf Freilandbedingungen übertragbar sind, sollte kontrollierte Flutung von Feuchtgebieten den Fortbestand von Regenwurmpopulationen gewährleisten, indem Überstauungen auf die Wintermonate beschränkt und die Wasserstände im Sommerhalbjahr erhöht werden. So würde für eine solide Nahrungsgrundlage für gefährdete Watvögel gesorgt und einer ungewünschten Nährstofffreisetzung ins Flusswasser vorgebeugt.

1. General Introduction

„ In our current state we run the same risk of Kipling's blind men trying to describe an elephant: our knowledge of the parts may obscure our understanding of the whole "

William B. Bowden (1987) about biogeochemical research in wetland ecosystems

1.1 Wetlands

Ecosystem research in terrestrial and aquatic systems has developed as two separate subdisciplines. Mainly soil ecology and limnology deal with nutrient dynamics in wetlands, but separately. To be able to deal with biogeochemical phenomena across large landscapes those subdisciplines have to be merged (Grimm et al. 2003). The present study concentrates on the interface of a terrestrial and an aquatic system, one of the so-called biogeochemical "hot spots". These are areas were hydrological flow paths converge with substrate containing complementary reactants. As a consequence, these hot spots show disproportionately high reaction rates relative to the surrounding area (McClain et al. 2003).

The RAMSAR convention (1971, in Middleton 1999) defined wetlands as "areas of marsh, fen, peatland or water, whether natural or artificial, permanent or temporary, with water that is static or flowing, fresh, brackish, or salt, including areas of marine water the depth of which at low tide does not exceed six meters". A broader definition is given by Cowardin et al. (1979): *"Lands transitional between terrestrial and aquatic systems where the water table is usually at or near the surface or the land is covered by shallow water"*. Both definitions can be applied to the sites studied in this thesis as they are situated in river floodplains (see Chapter 3).

Wetlands provide different ecosystem services. Their importance for the protection of species has been known for a long time, while other services were discovered only recently (Hoffmeister 2003). The retention of excessive water in wetlands prevents inundations in settled areas. By flooding nutrient-poor grassland, the concentration of undesired nutrients and pollutants in the flood water can be reduced (Haslam et al. 1998). Wetlands are therefore called the "kidneys of the landscape". Although national and European laws clearly recommend the protection and restoration of wetlands, the acceptance of their "public services" is low. The destruction of wetlands by soil sealing, construction activities and pollution are going on in most places (Hoffmeister 2003).

Under natural conditions, floodplain meadows in Europe are irregularly flooded during winter. Controlled inundation in the floodplains of the rivers Rhine, Weser and other rivers is practised at a large scale. Furthermore, wetland ecosystems are promoted as a habitat for special plant communities and birds. Soil invertebrates are an important food resource for wading birds, and high invertebrate densities, especially of earthworms and large insect larvae, are a desired goal of wetland management (Brandsma 1997, Ausden et al. 2001).

1.2 Earthworms and enchytraeids

Earthworms and potworms (Annelida, Oligochaeta: Lumbricidae, Enchytraeidae) belong to the dominant soil invertebrate groups in wetlands. Most of them are soil-dwelling and prefer humid conditions. Only some species, such as the earthworm *Eiseniella tetraedra* (Sims and Gerard 1985) and the enchytraeid *Marionina riparia* (U. Graefe, pers. comm.) are also frequently found underwater (e.g. in ditches). Other species are able to cope with inundations, but survival depends on a species and its adaptations, as well as on the duration of flooding and the oxygen saturation of the water. Details on survival strategies are discussed in Chapter 4. Earthworm abundance is normally reduced by extensive flooding (Volz 1976, Ekschmitt 1991, Graefe 1998, Faber et al. 2000, Römbke et al. 2002, Tabeling and Düttmann 2002). Little is known about enchytraeids in flooded sites. In regularly winter-flooded sites (Graefe 1998) as well as in bogs (Healy 1987) they occur as a very diverse group. However, compared to well-drained soils their density is rather low in very wet sites (Beylich and Graefe 2002).

Both groups, earthworms and enchytraeids, are decomposers. They take up and break down plant material. They mix it with soil and enrich it with substrate for micro-organisms (mucus, faeces). As a consequence, they have a large, mostly indirect impact on nutrient turnover. The mobilising influence of oligochaetes on nutrients in terrestrial systems is well known (earthworms: Sharpley and Syers 1977; Ruz-Jerez et al. 1992, Binet and Trehen 1992; Scheu 1993, Schrader et al. 1997, Haynes et al. 2003; enchytraeids: Wolters 1988, Williams and Griffiths 1989, Sulkava et al. 1996, Briones et al. 1998). In wetlands the mobility of nutrients is much higher, as water to solute them is present most of the time. As a consequence, mobilising soil

2

invertebrate effects could be even more pronounced and contribute to the eutrophication of river waters.

Also an immobilising influence of annelid worms on nutrients is possible and would be desired as a contribution to the purification of river water (Haslam et al. 1998). Annelids foster the growth of micro-organisms, and these use nutrients for building up their biomass (earthworms: Pedersen and Hendriksen 1993; Daniel and Anderson 1992; Tiwari and Mishra 1993; enchytraeids: Hedlund and Augustsson 1995). Thus, an indirect immobilising effect can be assumed and has been observed in some cases (e.g. Nagel et al. 1995). Additionally, as the earthworms' guts and casts are denitrification hot spots (Svensson et al. 1986; Scheu 1987; Parkin and Berry 1994; Karsten 1997), nitrogen could be removed from soil and water via this path.

Until now, the role of invertebrates in a terrestrial-aquatic environment has only been studied through lake sediments. Sediment-dwelling macrozoobenthos (i.e. tubificids and/or chironomid larvae) can have both a mobilising (Fukuhara and Sakamoto 1987) as well as an immobilising effect on nutrients (Lewandowski and Hupfer, in press). Terrestrial invertebrates have been neglected in previous studies.

1.3 Aims and outline

The aim of the present study was to determine the role of soil macrofauna in terrestrial-aquatic nutrient exchange (nitrogen, phosphorus) in flooded grassland sites with soil types mainly differing in SOM content. To approach this research question, a literature review, a field survey extending over two years as well as four different laboratory and one field experiment were carried out.

In two general preparing chapters, dynamics of nitrogen and phosphorus in wetlands are summarised (Chapter 2) and the three study sites are characterised (Chapter 3). In preparing the experiments for the main research aim, two other problems were addressed:

1. What influence do intensity, frequency, and time of flooding have on soil invertebrate populations, their abundance, biomass, diversity and species composition? How do soil invertebrates react towards flooding, and what are the most successful survival strategies?

The question was first addressed in a literature review about soil invertebrates (gastropods, earthworms, enchytraeids, insect larvae, isopods, diplopods and chilopods) in flooded grassland (Chapter 4). Then it was applied to the soil fauna of three floodplain study sites in northern Germany. In a first field survey, earthworms (Lumbricidae) and potworms (Enchytraeidae) were found to be the dominating groups in these sites. It was decided to concentrate all further research, including the laboratory experiments, on these two annelid groups. To be able to simulate natural abundances of annelids in these experiments, they were assessed in a field investigation (Chapter 5) with the research question:

2. How do annelid populations react to changing environmental conditions, with special respect to hydrological extremes (floods and drought) in wetlands?

Originally, three to five sampling dates spread seasonally over one year were planned. However, because of the extraordinary hydrological conditions (next to the regular winter inundations there was a summer flood in two of the sites and an exceptional drought in the subsequent summer), ten censuses were carried out in the course of two years.

Furthermore, data from these field investigations were comprehensive enough to determine the efficiency of the mustard extraction method for earthworms. Chapter 6 addresses the question:

3. Is the mustard extraction method suitable to sample different earthworm species in different soils under different hydrological conditions?

The first laboratory experiment combined in its design all three sites and both annelid groups. Mesocosms with undisturbed soil samples were inoculated with annelid worms and flooded with water from the respective river. Nutrient concentrations were compared with an animal-free control (Chapter 7), addressing the following question:

4. Are nutrients mobilised or immobilised in the presence of annelid worms in flooded soils?

From terrestrial systems, considerable effects of annelids on nutrient dynamics in both directions, mobilising and immobilising is known (see above). Thus, they could also play an important role in nutrient exchange between soil and water during flood events.

As the soils of the three study sites (marsh soil, peat soil, gley soil) differed in the content of soil organic matter (SOM) and contamination, the question was raised:

5. Are there differences concerning the role of annelids in nutrient release between different soils?

As the most distinct results in this first experiment were obtained for the peat soil, further experiments were restricted to this soil type.

In studies on fungal mycelium and microbial activity, a density-dependent effect of enchytraeids has been observed by Hedlund and Augustsson (1995). Such an effect is possible also for enchytraeids that occur in very high, but also variable densities from few individuals m-² up to several 10.000 individuals m-² (Petersen and Luxton 1982). Sulkava et al. (1996) found a positive correlation between enchytraeid biomass and the amount of NH_4^+ in soil.

To assess the density-dependence of their effect on nutrient release, another comprehensive microcosm experiment (Chapter 8) was designed to answer the question:

6. Is the effect of enchytraeids on nutrient dynamics in flooded soils density-dependent?

Additionally, two smaller experiments with earthworms were carried out to illuminate special characteristics of their role in nutrient release (Chapter 9):

7. At which moment in nutrient release does the presence of earthworms play a role?

When soil contacts with river water at the beginning of an inundation, this is a biogeochemical "hot moment" (McClain et al. 2003) in which normally a high amount of nutrients is released. The development of nutrient concentrations in the floodwater of peat soil microcosms was measured in the presence and absence of earthworms during the first 24 hours of flooding.

8. Which role does the dead tissue of earthworms play in nutrient mobilisation in flooded soils?

As annelid populations are strongly reduced by flooding (Ekschmitt 1991, Pizl 1999, Chapter 5), the role of their rapidly decaying dead tissue in nutrient release has to be taken into account.

In an attempt to support some of these laboratory results under natural conditions, a field experiment was set up in September 2004. Unfortunately, the flood simulation in soil mesocosms on the sites failed. The results concerning general nutrient release as well as a possible improvement of the experimental design are discussed in Chapter 10.

In the General Discussion (Chapter 11), nutrient release and annelid effects are compared throughout the different experiments. The focus lies on the relevance of the results for water management.

2. Nutrient dynamics in wetlands

For a better understanding of annelid effects on nitrogen and phosphorus dynamics, it is necessary to know which pathways these nutrients may take, under which environmental conditions they occur in which form and which micro-organisms are involved in their transformation.

2.1 Nitrogen cycling and transformation in wetlands

The three inorganic nitrogen ions (ammonium, nitrite, and nitrate) are water-soluble salts. Consequently, they are distributed in dilute aqueous solution throughout the ecosphere.

The cycling of nitrogen through the biosphere largely determines ecological productivity in aquatic and terrestrial ecosystems.

2.1.1 Nitrogen pools

In wetlands, soils and sediments constitute the largest single pool of nitrogen (100 to several 1000 g N m^{-2}), followed by plants and available inorganic nitrogen (Bowden 1987). Living and dead organic matter represent smaller reservoirs from where nitrogen can be actively released, just like from water. At least in temperate climates, more stable nitrogen reservoirs occur in the stabilised soil organic matter (or humus) where a slow mineralisation makes nitrogen available for uptake by living organisms (Atlas and Bartha 1998). During the vegetation period, plants are a sink for N (and P) while organic material becomes a source of nutrients when it is mineralised. Mineralisation is slower when conditions are water-saturated (Meuleman 1999). Kalbitz et al. (2002) measured high N mineralisation peaks and a reinforced peat decomposition after a lowering of the groundwater table in summer and a slow re-wetting in autumn. This is how increasing groundwater tables – and flooding - may partly decrease the mineral nitrogen content of the topsoil, when nitrogen that was mineralised during a dryer period becomes available to ground- or surface water. Wetlands often function as transformers of nutrients from organic to inorganic forms, i.e. the input is just as high as the output (Mitsch and Gosselink 1993, Hefting 2003). N-mineralisation in wetlands strongly depends on the substrate quality of the soils as well as on oxygen supply. In a study by Updegraff et al. (1995), N-mineralisation was much higher in a sedge meadow than in a spruce-*Sphagnum*-bog. Surface peat in the latter had the highest nitrogen mineralization rates under aerobic conditions. Under anaerobic conditions, methane was produced.

Most of the inorganic nitrogen in wetland soils occurs in the form of ammonium. Nitrate is generally scarce or absent in wetland soils (Bowden 1987).

2.1.2 Nitrogen input

Nitrogen mainly enters the biosphere by N_2-fixation and anthropogenic input (via atmospheric or hydrological deposition). Less frequent processes, such as volcanic activity, ionizing radiation and electrical discharges supply additional combined nitrogen to the atmosphere. This nitrogen becomes available to the biosphere when it is washed down with precipitation (Atlas and Bartha 1998, p. 415). In floodplains, river water can be an important source of nutrients for the flooded wetland.

The atmospheric deposition can be an important source of new nitrogen, even in hydrologically open wetlands, i.e. not only in ombrotrophic (i.e. rain-fed) peatlands (Bowden 1987). This may have diverse effects, such as eutrophication, acidification, and a change in species composition (Bobbink and Lamers 2002). Wet deposition by rainfall adds roughly 0.5 to 1 g N m^2 yr^{-1}, depending on the local source areas near the wetland. The hydrological input of N is even more dependent on the types of activities in upland areas. These activities associated to agriculture, industry, mining or human settlements do not only determine the amount of N but also whether it is added as nitrate, ammonium, or organic N (Bowden 1987).

2.1.3 Nitrogen export

The fraction of nitrogen that is exported by effluent water is much smaller than the potentially mineralisable amount. In general, the export of N from wetlands is small, and in many cases, wetlands may be net sinks for hydrologically transported N (Bowden 1987). Other wetlands may be nitrogen sinks only seasonally, taking up nitrogen during the growing season but remobilising it during the off-season (Simpson et al. 1978; Hoffmann et al. 2003).

Leaching is the rapid loss of water-soluble nitrogen compounds. Microbes that rapidly colonise dying (plant) tissues can reduce leaching losses. In marshes, fresh, aerobic litter of plants such as *Typha, Carex* or *Calamagrostis* sp. persists on the surface all winter. This litter acts like a protecting cap that prevents nitrogen loss from the pore water of sediments - which normally has a high N-concentration - to the river water, which has a much lower N-concentration. The litter may even extract nitrogen from the river water when the wetland is flooded. Thus, a wetland as a whole immobilises nitrogen even when the plants are not actively assimilating nitrogen. This nitrogen stock acts as a buffer during short-term disruptions in N supply, such as might occur during drought. However, when the litter is compacted to anaerobic peat, it becomes a net source of nitrogen available to plants. Export then takes place as denitrification (formation of volatile NO_x gases; Bowden 1987).

In freshwater wetlands, there is generally more nitrogen cycling in the system than leaving or entering the system. The following paragraphs describe the parts of nitrogen cycling in which soil animals are involved (see also Fig. 2.1).

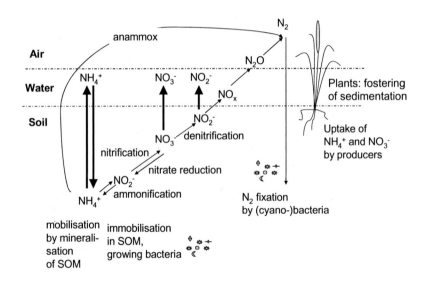

Figure 2.1: Nitrogen cycling in wetlands. Simplified scheme developed after Fritsche 1998, Bick 1999, Schultz 2000, Wienk et al. 2000, Richardson and Schlesinger 2004

2.1.4 Nitrification and other ammonium oxidation pathways

During nitrification, ammonium ions are oxidised to nitrite and further to nitrate:

Nitrification: $2 NH_4^+ + 3 O_2 => 2 NO_2^- + 4 H^+ + 2 H_2O => 2 NO_3^- + 8 H_2$

Nitrification occurs readily in neutral, well-drained soils, while it is inhibited in anaerobic or highly acidic soils. Normally, the formation of nitrite and nitrate are closely coupled. The second step of nitrification, the transformation of NO_2^- to NO_3^-, is oxygen-dependent. When oxygen deficiency occurs and / or the production of nitrous acid (HNO_2) lowers the pH of the environment, nitrite is accumulated.

The oxidation of ammonium into nitrite and nitrate results in a change from a positive to a negative charge. Positively charged ions tend to be bound by negatively charged clay particles in soil, but negatively charged ions migrate in the soil water. Thus, nitrification can be a nitrogen-mobilisation process within soil habitats. The products, nitrite and nitrate, can also be readily leached from the soil column into the groundwater, and fixed forms of nitrogen are lost from the soil (Atlas and Bartha 1998).

In soil, *Nitrosomonas* is the dominant genus oxidising ammonia to nitrite (next to *Nitrosospira, Nitrosococcus, Nitrosolobus*, and *Nitrosovibrio*), while nitrite is turned into nitrate by *Nitrobacter* (next to *Nitrospira, Nitrospina,* and *Nitrococcus*).

Nitrification may be autotrophic ($NH_4^+ => NO_3^-$) or heterotrophic (reduced organic N $=>$ oxidised N compounds). In wetlands, the lack of available O_2, the acid conditions, the competition from plants, allelopathy, and a lack of available phosphorus normally limit autotrophic nitrification (Bowden 1987). Heterotrophic nitrification is less well studied. The same is true for the recently discovered anaerobic ammonium oxidation (anammox) which is an alternative pathway of ammonium transformation used in the purification of sewage (www.anammox.com, October 2004). This could play an important role in anaerobic wetland soils. Anammox probably transforms 65% of the total nitrogen in wetlands (Richardson and Schlesinger 2004).

Anammox: $NH_4^+ + NO_2^- = N_2 + 2 H_2O$

2.1.5 Ammonification

In living and dead organic matter, nitrogen occurs predominantly in the reduced amino form. This organic nitrogen is converted to ammonia (NH_3) during ammonification.

Several plants, animals, and microorganisms are capable of this organic nitrogen mineralisation. In alkaline environments, ammonia is released into the atmosphere where it is relatively inaccessible to biological systems. In neutral and acidic aqueous environments, ammonia is transformed into ammonium ions (NH_4^+) which can be assmimilated by plants and microorganisms. They are incorporated into amino acids and other nitrogen-containing biochemicals (Atlas and Bartha 1998).

Ammonium not taken up by plants or tightly bound to sediments may be nitrified (Bowden 1987). The vicinity of aerobic and anaerobic areas in wetlands is crucial for the conservation of N within the system: The diffusion of ammonium from the anaerobic to the aerobic layer happens slowly (0.059 to 0.216 cm^2 day^{-1}), just as its oxidation in the aerobic layer (Reddy et al. 1980).

2.1.6 Denitrification

During denitrification, organic matter is oxidised, while nitrate is reduced completely through nitrite to nitric oxide (NO_x) and nitrous oxide (N_2O) and further to molecular nitrogen.

Denitrification: $NO_3^- => NO_2^- => NO => N_2O => N_2$

Heterotrophic denitrification: $5(CH_2O) + 4NO_3^- + 4H^+ => 5CO_2 + 2N_2 + 7H_2O$

On this path, molecular nitrogen returns to the atmosphere. Facultative anaerobic microorganisms in soils, sediments, and guts of higher organisms perform denitrification (Atlas and Bartha 1998). The surface soil of three soils in a riparian environment (coniferous forest peat, mixed forest, and marsh soil) had a higher denitrification-potential than subsurface layers (0.8 – 1.4 m; Hill and Cardaci 2004).

The proportion of the different denitrification products depends on the involved organisms as well as on environmental conditions. The lower the pH of the habitat, the greater the proportions of nitrous oxide formed. Under waterlogged conditions, oxygen tensions are usually low. This favours denitrification and consequently nitrite and nitrate are usually rare (Bowden 1987). Additionally, plants may enhance soil N transformation and N_2O emission under wet soil conditions (pF = 2; Augustin et al. 1997).

In wetland soils and sediments, nitrification and denitrification often occur in close proximity, and a great part of NO_3^- formed by nitrification diffuses to the anaerobic denitrification zone where it is reduced to N_2 (Kester et al. 1997, Trepel and Kluge 2002, Well et al. 2002, Hefting 2003). However, nitrification and denitrification both stay far below the potential level at optimal conditions, i.e. in drained soils or with fluctuating water levels (Bowden 1987).

Using the ^{15}N-technique, the authors found out that nitrate was transformed by denitrification (to N_2) and by dissimilative reduction (to ammonium). However, in the field, ammonium concentrations in river water also tended to be lower after the inundation. Thus, it is not evident that nitrate is reduced via this path in the field (Hoffmann et al. 2003).

2.1.7 Other nitrate reduction pathways

The reammonification or dissimilatory nitrate reduction conserves nitrogen as ammonium within the system. Probably, this pathway is only quantitatively significant in salt marshes (due to the SO_4^{2-} reduction couple). However, it is also possible that a sufficient supply of NO_3^- (as it might occur in freshwater wetlands) leads to a considerable amount of reammonification (Bowden 1987).

Re-ammonification: $NO_3^- + 2H^+ + 8[H] => NH_4^+ + 3H_2O$

In assimilatory nitrate reduction, nitrate ions are reduced to ammonia by enzymatic reactions of a variety of organisms (bacteria, fungi, and algae) and consequently incorporated into organic matter. In the absence of oxygen, nitrate ions can act as terminal electron acceptors in nitrate respiration or dissimilatory nitrate reduction. While organic matter is oxidised, nitrate is converted to various reduced products, mainly nitrite, via facultatively anaerobic bacteria (e.g. *Alcaligenes, Escherichia, Aeromonas, Enterobacter, Bacilllus, Flavobacterium, Nocardia, Spirillum, Staphylococcus,* and *Vibrio*). Some of these organisms will reduce the excreted nitrite via hydroxylamine to ammonium. This nitrate-ammonification plays an important role in stagnant water. This process does not result in any gaseous nitrogen products and ammonia does not inhibit it (as it does in assimilatory nitrate reduction).

Therefore, ammonium ions can accumulate in relatively high concentrations (Atlas and Bartha 1998).

2.2 Phosphorus dynamics in wetlands

In contrast to the complex oxido-reductive cycling of nitrogen, phosphorus only exists in one stable valence. In microbial cycling, the oxidation state of phosphorus is generally not altered (i.e. there are no redox-reactions). It is simply a transfer from insoluble, immobilised organic forms to soluble, mobile inorganic compounds, balanced by mineralisation and sedimentation (Atlas and Bartha 1998). However, phosphorus occurs in various forms, and the determination of certain fractions is technically complicated (Richardson 1999).

Phosphorus is an essential element for biological systems where it occurs in phosphate esters and nucleic acids (Atlas and Bartha 1998, Kölle 2001). It is normally not abundant in the environment and often limits microbial growth. In the presence of bivalent metals (Ca^{2+} and Mg^{2+}) and ferric ions (Fe^{3+}), it has the tendency to precipitate at neutral to alkaline pH, e.g. due to the formation of Fe(III)hydroxides at redox interfaces. Most phosphorus losses are due to fixation in sediments of rivers and lakes (Atlas and Bartha 1998). Higher water tables on peat soil may increase phosphorus contents in the groundwater due to a decreased redox potential, which again may increase phosphorus solubility and intensify leaching of dissolved organic matter (Kalbitz et al. 2002).

Chemical processes primarily control P-fluxes from peat soil to water (Richardson and Marshall 1986). The phosphate dilution normally goes along with the production of organic acids and as a consequence, a decrease of the water pH. Under anaerobic conditions, such as during the flooding of soils, even the insoluble ferric phosphates can be reduced by microorganisms that reduce ferric ions to ferrous ions. As a result, phosphorus is released into the water (Atlas and Bartha 1998). Zak et al. (2004) measured high phosphorus concentrations in the anaerobic pore water of rewetted peatlands. However, the adjacent surface waters were not eutrophicated, as there was a strong retention of P by the precipitation of Fe(III) oxyhydroxides. The authors only expect a P export from such peat soils when the Fe/P ratio is smaller than 3. Soluble inorganic forms of phosphorus are normally assimilated and thus immobilised by plants and microorganisms (Atlas and Bartha 1998).

3. Study sites and rivers

As the role of soil fauna in different soil types was of special interest for this study, sites contrasting especially in the SOM content of their soils were selected. Figure 3.2 shows the geographic situation of the three study sites in Northern Germany. An organic soil in a nature reserve in Bremen (peat soil) was available for scientific study. With an official soil research study site of the State Lower Saxony in Gorleben, a site with mineral soil (gley soil) with various contaminations from the river water was also available. Actually, a comparing site without contamination would have given the ideal conditions to investigate the influence of this special characteristics on ecosystem processes. Unfortunately no such site was available. Instead, a marsh soil site was chosen. It takes an intermediate position concerning SOM content and it is more representative for the region around Bremen.

All three sites are regularly winter-flooded meadows along rivers in Northern Germany, each differing in soil type (river Wümme: peat, river Ochtum: marsh soil; river Elbe: contaminated gley soil). Site characteristics are shown in Table 3.1.

Table 3.1: Vegetation and soil characteristics of the study sites

Main data sources: [a] Erber 1998; [b] B. Olbrich, pers. comm.; [c] B. Schuster, unpublished data; [d] Janhoff 1992; [e] G. Oertel, pers. comm., [f] Kleefisch and Knes 1997; [g] own measurements / observations; [h] www.wetteronline.de/reuro.htm, 21.10.2004.

Study site	Nature reserve "Ochtumpolder bei Brokhuchting", Bremen	Nature reserve "Borgfelder Wümmewiesen, Bremen	Soil monitoring site of the State Lower Saxony, Gorleben/Elbe
Soil type	marsh soil[a]	peat (10-15 cm) on river sand[d]	river gley soil[f]
pH ($CaCl_2$)	4.8 [a]	4.6 [c, d]	5.5[f]
N_t (%)	0.7[a]	2.3 [c, d]	0.7[f]
C_{org} (%)	6.2[a]	29.1 [d]	9.5 [f]
C/N	9.1[a]	14.8 [c, d]	13.4[f]
SOM (%)	> 10[a]	27.8 [c, d]	9.3[f]
Field capacity (vol.-%)	64[a]	67 [d]	64[f]
Water holding capacity (mass %)	64[a]	76[g]	52[f]
Special characteristics	oxide concretions[a]	peat mixed with river sediments[d]	contamination (see text)[f]
River (tributary to stream)	Ochtum (Weser)	Wümme (Weser)	Elbe
Climate	oceanic	oceanic	subcontinental
Water management, inundation interval	backwater flooding every winter from December to February[b]	backwater flooding in winter, episodic summer floods at high water levels[d]	no management, natural inundations every second winter, episodic summer floods[f]
Inundations before and during the study period (exact dates only when observed by the authors)	Dec. 01 to Feb. 02, (rainwater 07-24 to 08-02-2002), 12-15-02 to 02-15-03[g]	Sept. 01, Nov. 01 to Apr 02, 07-20 to 08-21-2002, 10-22-02 to 03-15-03[g]	Nov. 01 to Feb. 02, 08-18 to 09-11-2002, Nov 02 to Feb. 03[g]
Plant association	Phalaridetum arundinaceae[a, g]	Caricetum gracilis (Calthion)[d, g]	Phalaridetum arundinaceae, Glycerietum fluitantis[f, g]
Root horizon	10 cm[a]	13 cm[d]	7 cm[f]

Table 3.1: Vegetation and soil characteristics of the study sites (continued)

Study site		Nature reserve "Ochtumpolder bei Brokhuchting", Bremen	Nature reserve "Borgfelder Wümmewiesen, Bremen	Soil monitoring site of the State Lower Saxony, Gorleben/Elbe
Agricultural use		mowing 1-2 times/year; sometimes pasture in autumn (not in years of study) [b]	mowing 1-2 times/year [e]	mowing 2-3 times/year [f]
precipita-tion (mm y^{-1}); long-term mean	2002	1070 [h]	1070 [h]	852 [h]
	2003	607 [h]	607 [h]	427 [h]
sun hours 2002		1523 [h]	1523 [h]	1501 [h]
sun hours 2003		1885 [h]	1885 [h]	2005 [h]

3.1 Marsh soil site „Ochtumniederung bei Brokhuchting"

The marsh soil site is situated in the nature reserve "Ochtumniederung bei Brokhuchting" in the south-west of Bremen (Figure 3.3). It is protected according to the European bird directive and has also been proposed as FFH (Flora-Fauna-Habitat) area (Senator für Bau, Umwelt und Verkehr 2004). The "Ochtumniederung" is a polder, i.e. an inundated area surrounded by dykes, in a traditional cultural landscape in the southwest of Bremen called "Niedervieland". This landscape is characterised by a dense network of smaller and bigger ditches which convey the flood water to the meadows. The meadows are structured by a man-made micro-relief consisting of more shallow ditches, ensuring drainage of the ridges, even in very wet conditions. In winter the ridges are flooded, too. In early spring the reserve is frequented by various water birds (several duck and goose species) and waders. The meadow is mown once or twice a year. Occasionally, it is used as a horse pasture in autumn (B. Olbrich, pers. comm.). However, this was not the case in the years of study.

The river Ochtum is a direct tributary on the left side of the river Weser which flows into the North Sea at Bremerhaven. The Ochtum's catchment is situated near Bremen. As it is small and the water regime is regulated by a river flood barrier downstream, there are no natural inundations, not even after heavy rainfalls or spring tides of the North Sea. All inundations are brought about by water management and therefore they are predictable in the range of a few days. Controlled flooding normally starts in late October when the river Ochtum has enough water. The highest parts of the polder are flooded around Christmas. In the beginning of February, the water is allowed to leave the polder again (B. Olbrich, pers. comm.). However, stagnant water in depressions provides ideal conditions for birds to search for food until late spring (A. Schoppenhorst, pers. comm.).

3.2 Peat soil site "Borgfelder Wümmewiesen"

The peat soil site is situated in the core zone of the nature reserve "Borgfelder Wümmewiesen" in the north-east of Bremen (Figure 4). The reserve is protected according to the European bird directive and the FFH directive. At the same time, it is a reserve with "representative importance for the whole Federal Republic" ("Naturschutzvorhaben mit gesamtstaatlich repräsentativer Bedeutung", Senator für Bau, Umwelt und Verkehr 2004). An area of 677 ha is regularly winter-flooded by the river Wümme, a secondary tributary on the right side of the river Weser. The area is subject to hydrological management, although at high water levels flooding occurs naturally. The extended catchment area includes a large part of the nature conservation park "Lüneburger Heide". Intensive rainfall there leads to natural inundations, sometimes also in the warmer season. Depending on hydrological conditions, the meadow is mown once or twice a year.

There are breeding populations of corn crake (*Crex crex*), spotted crake (*Porzana porzana*), and curlew (*Numenius arquata*; W. Eikhorst, pers. comm.). Just like the Ochtum polder, the area is frequented by large populations of migrating water birds during the extended winter inundations that often last until late spring, with whooper swans (*Cygnus cygnus*) being the most prominent species. Waders such as lapwing (*Vanellus vanellus*), common snipe (*Gallinago gallinago*), redshank (*Tringa totanus*) and black-tailed godwit (*Limosa limosa*) also breed in this area. Bigger soil animals such as earthworms and *Tipula* larvae are an important food resource for them. In the previous decade, their breeding numbers were declining. This is due to predation of eggs and young birds by craws and foxes (A. Schoppenhorst, pers. comm.), but also a possible relation to declining earthworm populations can be considered (Göbel 2003).

As the study site is situated in the core zone of the nature reserve where no fertilisation is allowed, the N export during inundations is very low (about 6.4 kg N ha^{-1} year^{-1}). However, the N pool in this soil is remarkable (about 45.000 kg N ha^{-1} in the upper 90 cm). As nitrification is impeded in this acidic soil, the dominating end product of mineralisation is ammonium (Scheffer and Ausborn 1998).

The river Wümme has a good water quality (see Table 3.2) with only a low level of contamination. However, increased concentrations of cadmium are present in the sediment. Downstream from the nature reserve, nitrate values are somewhat lower than upstream. This is also the case in spring when the discharge is highest and nitrate concentrations are increased up to 3.6 mg l^{-1} upstream (2.2 mg l^{-1} downstream). In late summer and autumn, all nutrient concentrations show lowest values, and ammonium (< 0.05 mg l^{-1}) and nitrite (< 0.01 mg l^{-1}) are hardly measurable (Niedersächsisches Landesamt für Ökologie 2001).

Table 3.2: Water parameters of the river Wümme. Median of 12 measurements at monthly intervals in 1999. Upstream from the study site: Ottersberg / Wümme Nordarm; downstream from the study site: Truperdeich/Wümme. Source: Niedersächsisches Landesamt für Ökologie 2001.

parameter (mg l^{-1})	ortho- PO$_4$-P	total P	NH$_4$- N	NO$_2$- N	NO$_3$- N	total N	TOC	DOC	HCO$_3$	AOX
upstream	0.03	0.16	0.08	0.02	1.5	2.0	8.0	8	119	18
downstream	0.04	0.17	0.13	0.03	1.3	2.0	9.0	-	-	-

3.3 Gley soil site "Elbaue bei Gorleben"

The gley soil site is situated within the floodplain of the river Elbe discharging into the North Sea downstream of Hamburg) in the north-east of Lower Saxony. It is situated in the middle reaches of the Elbe stream, in a subcontinental climate, i.e. with less rainfall and more frequent hot summers than Bremen (Wetteronline.de 2004, review for the last twenty years).

The study site is a permanent soil monitoring site of the State Lower Saxony. Its exact situation is kept secret and cannot be shown in a map in this thesis. The site is a mowing meadow without any application of fertilizer (B. Kleefisch, pers. comm.). In the region around the study site, the river Elbe has a critical water quality with various pollutants introduced from industrialised regions upstream. The sediments, just like the soils in the regularly inundated wetlands along the river, have increased concentrations of cadmium, nickel, lead, zinc, mercury and copper (Niedersächsisches Landesamt für Ökologie 2001, Kleefisch and Knes 1997). The metabolic quotient in this soil is very high compared to uncontaminated reference sites, while the quotient C_{mic}/C_{org} is significantly lower, hinting at an acute as well as long-term pollution. The acute effect is a reduction of the carbon profit of microorganisms. However, it remains uncertain if the impeded microbial activity is due to these contaminations or rather to the high groundwater table (Höper and Kleefisch 2001).

Figure 3.1: Geographical situation of the three study sites in Northern Germany. Map taken from Diercke Weltatlas, 3rd edition 1992, p. 46, with kind permission of Westermann Schoolbook Publishers.

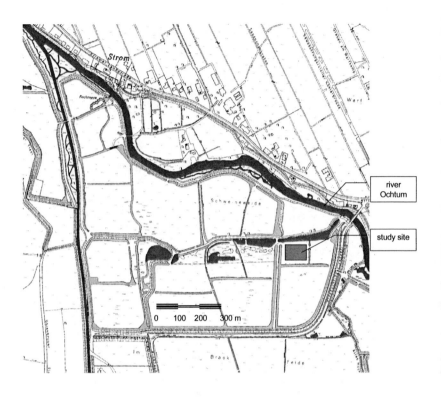

Figure 3.2: Situation of the marsh soil site in the nature reserve "Ochtumniederung bei Brokhuchting". Dark fields: water bodies (river Ochtum flowing to the north-west, side branches and ditches). Water cartography by B. Olbrich, BUND, on a map of the Niedersächsisches Landesvermessungsamt. With kind permission.

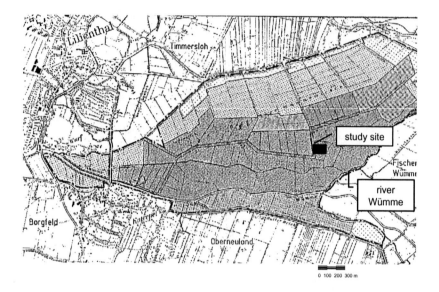

Figure 3.3: Situation of the peat soil study site in the nature reserve "Borgfelder Wümmewiesen". Darker grey: core zone of the reserve, lighter grey: buffer zone. Cartography by Planungsgruppe Grün, Bremen on a topographical map 1 : 25:000 (Niedersächsisches Landesverwaltungsamt - Landesvermessung 1990, with kind permission).

4. Terrestrial invertebrates in flooded grassland: a literature review

published as: Plum, N.M. (2005): Terrestrial invertebrates in flooded grassland: a literature review. *Wetlands* 25 (3), p. 721-737

Abstract: I reviewed information from 73 sources on soil animals in flooded grassland, focusing on soil macrofauna: snails and slugs (Gastropoda), earthworms (Lumbricidae), potworms (Enchytraeidae), woodlice (Isopoda), millipedes (Chilopoda, Diplopoda), Diptera, and other insect larvae. While the database on earthworms was rather comprehensive, studies on all other soil fauna groups, especially regarding their role in wetlands were few and showed major research needs. The survival strategies of the groups were compared systematically. Annelids and insect larvae had the best physiological adaptations, other groups only reacted with evasion by active or passive movement and by recolonisation and reproduction from resistant stages. There were no typical "wetland species" in soil macrofauna, only tolerant hygrophilous species. Flooding of grassland immediately reduced diversity, abundance, and biomass of all groups of soil macrofauna. Their community structure was altered, and well-adapted species, often wide-spread opportunists, became more abundant while others disappeared. The effects increased with the duration of flooding and rising temperature but were usually compensated for during the next soil-dry period. A meta-analysis classified sites according to duration, frequency, and seasonality of inundation. In general, species numbers and abundances of earthworms, woodlice, and millipedes tended to be lower in frequently and/or extensively flooded sites. Only gastropods are favored by moderate winter flooding. In bogs, even when they are waterlogged the entire year, species numbers are distinctly higher than the most frequently flooded sites. The impact of episodic summer flooding events is transitional and less pronounced than that of regular winter flooding. Earthworms re-establish soil structure after flooding, and they are, next to dipteran larvae, an important prey for wetland birds. Slow, moderate flooding in winter, waterlogging in spring, and a landscape mosaic with non-flooded refuge sites is recommended for water management policies favorable for both soil fauna and wetland birds.

Key Words: literature review, flooded grassland, flood tolerance, terrestrial invertebrates, soil macrofauna, Gastropoda, Lumbricidae, Enchytraeidae, Isopoda, Chilopoda, Diplopoda, Diptera, Coleoptera, Europe

4.1. Introduction

Since wetland ecosystems consist of two subsystems, an aquatic and a terrestrial one, a comprehensive evaluation of impacts should consider both (Grimm et al. 2003). However, most recent studies on wetland invertebrates have focused on the drought tolerance of aquatic species. This review summarizes how large terrestrial invertebrates tolerate flooding. The results of 76 recent studies, most of them from European countries, are presented. Several of these studies were only published in German, Dutch, French, Czech, or Slovakian. The data about population dynamics of terrestrial invertebrates are used for a meta-analysis of the factor "flooding" on their species numbers, abundance, and biomass.

The review was restricted to soil macrofauna because of their important effect on physical soil properties and organic matter decomposition. The majority of soil macrofauna studies in wetlands have focused on earthworms (Lumbricidae), which are known to be ecosystem engineers and usually dominate soil fauna biomass. There are a number of studies on isopods, diplopods, and chilopods, which often have been treated together, whereas data on dipteran and coleopteran larvae, as well as on gastropods (especially slugs), are scarce. Enchytraeids as representatives of the soil mesofauna were included in this review because they are known to be most abundant in wet, especially acidic soils and locally can surpass all other taxa of soil macrofauna, if not in biomass, then in respiration rates (Curry 1994).

After a short overview of the threats to soil animals caused by flooding and the principal survival strategies, their occurrence and population dynamics in flooded grassland are compared. Sampling data are used for a meta-analysis of species numbers, abundance, and biomass of soil invertebrates at different inundation intensities. The results are summarized and discussed against the background of the survival strategies. The role of terrestrial invertebrates in wetland soils, as far as research has been performed, is shown, and recommendations for wetland management are given.

4.1.1 Criteria for the selection of references

Databases available on the internet (especially the ISI Web of Science) were searched for relevant literature, preferring recent articles. The terms soil (fauna), invertebrates, terrestrial-aquatic, flood(ing), inundation, floodplain, wetland, and water management were entered in English, French, German, and Dutch in all possible combinations. Additionally, the scientific names of the studied soil animal groups were used for the search: Bibionidae, Chilopoda, Diplopoda, Diptera, Enchytraeidae, Gastropoda, Isopoda, Lumbricidae, Mollusca, Myriapoda, and Tipulidae.

Reports from German and Dutch research institutions, departments, and planning offices were integrated if standard methods were used for sampling soil fauna. The good availability of (unpublished) data from Western and Central Europe implicates a focus on this region. Necessary additional information, if not mentioned in the articles or reports, was collected by personal communication with the authors.

4.1.2 Study sites

The selection of study sites analyzed here is based on the rather broad definition of wetlands given by Cowardin et al. (1979): *"Lands transitional between terrestrial and aquatic systems where the water table is usually at or near the surface or the land is covered by shallow water."* The focus lies on flooded grassland that is natural (reeds, high sedges, herbaceous fallows) or used as meadow or pasture. Some observations on survival strategies of soil animals originate from wetland forests. Sampling results from non-flooded sites were taken into account if they were adjacent to flooded grassland and, thus, could be used for comparison. Artificially waterlogged fens and bogs, remote from rivers, were also included but treated separately. The studies containing information on species composition, abundance, and biomass of soil macrofauna and/or enchytraeids aimed at:

- characterizing the soil fauna of wet and flooded soils in general (Borcherding 1889, Volz 1976, Schröder 1980, Rusek 1984, Emmerling 1993, Dohle et al. 1999, Zerm 1999);
- establishing a system of decomposer communities (Graefe 1998, Graefe and Beylich 1999, Beylich and Graefe 2002);

- determining the suitability of soil organisms for indication of soil quality (Römbke et al. 2002), disturbance (Frouz 1999) or "naturalness" of floodplain ecosystems (Spang 1996)

- determining the effects of natural floods, especially in the warmer season (Pizl and Tajovský 1998, Tajovský 1998, Cejka 1999, 2004);

- providing an inventory of soil fauna species before flooding was (re-)introduced and/or for monitoring developments afterwards:
    - in natural river floodplains (Bolte and Moritz 1987, Ekschmitt 1991, Rodieck et al. 1992, Dahl et al. 1993, Handke et al. 1999, Hansen and Castelle 1999),
    - in artificially lowered floodplain meadows where the loam layer was removed down to the lower-lying sand to give river floods more space (Faber et al. 1999), and
    - in drained (and re-wetted) fens and bogs (Frouz and Syrovátka 1995, Keplin et al. 1995, Helling and Kämmerer 1998, Keplin and Broll 2002);

- comparing sites with different flooding regimes to evaluate the suitability of water management practices for wetland-related birds feeding on soil organisms (Brandsma 1992, 1997, 2002, Meenken 1999, Tabeling and Düttmann 2002, Faida et al. 2003, Göbel 2003), combined with field experiments (Ekschmitt 1991: cages to exclude birds; Ausden et al. 2001: flooding of soil samples to investigate reactions and survival of earthworms); and

- assessing the accumulation of contaminants from river sediments in soil animals (Ma et al. 1997, Weigmann and Schumann 1999).

4.2 Survival strategies of soil animals in flooded grassland

The factors potentially threatening inundated soil animals are respiration problems when the entire body is surrounded by water and the oxygen content in the water is low; swelling; being moved out of the habitat by flowing water, and being affected by toxic substances that are formed in flooded soils or by contaminants from the river water. The oxygen content of the flood water, depending on temperature, movement, and water exchange, is the most important factor, implicating several changes in environmental conditions. With the depletion of oxygen, the concentration of CO_2 rises. Toxic substances such as butyric acid, acetic acid, and methane can originate from decomposed plant material. If sulfates are present, hydrogen sulfide can also occur (Topp 1981, Mather and Christensen 1988). When flooding is introduced in formerly fertilized grassland, the decaying litter of flooding-intolerant grass species such as *Lolium perenne* L. lead to a quick depletion of oxygen (Ausden et al. 2001). Regular inundation results in a change of the plant communities from nutrient-rich meadow grasses towards reed species (*Phalaris arundinacea* L., *Deschampsia cespitosa* (L.) P.B., *Carex*-spp.), which normally have a higher C/N-ratio and decompose slowly. The food quality for soil animals thus deteriorates. In vegetation gaps where litter and growing plants are lacking after inundation, epigeic (i.e., surface-dwelling) species like the earthworm *Lumbricus rubellus* as well as fleeing endogeic (i.e., soil-dwelling) animals are endangered by higher soil temperatures, UV light (causing excitement, then paralysis, Graff 1983), and birds feeding on them (Erber et al. 2002). Besides, changes in environmental conditions due to flooding can have indirect negative effects on soil animals. Flooding can result in soil compaction, consolidation, and loss of soil structure. Soil cavities, the habitat of soil invertebrates, are scarce in regularly flooded soils, where they easily become blocked by sediments. Air and water movement through the soil is impaired, and the movement of soil animals is impeded (Ausden et al. 2001). Not only flooding, but also summer drought and the intense alternation of both in grassland soil, is a stress factor, which in continental regions is more marked than in oceanic. Following the respective precipitation maximum, spring flooding is more common in oceanic regions and autumn flooding in continental areas. (Schultz 2000). Soil animals are thus affected at different individual and population development stages. Normally, the inundation tolerance of arthropods is higher than their tolerance towards drought (Hildebrandt 1995a).

Table 4.1: Flooding survival strategies of different soil animal taxa

Taxa	horizontal migration	vertical migration	physiolog. adaptation	pheno-logy	repro-duction
Gastropoda	x	x			x
Lumbricidae	x	x	x	x	x
Enchytraeidae		x	x	x	x
Isopoda		x			x
Chilopoda				x	x
Diplopoda		x			
Insect larvae	x	x	x	x	x

The ability of soil animals to survive flooding depends on their behavioral, morphological, and physiological properties. Table 4.1 summarizes the different possible survival strategies of the studied soil animal groups. Earthworms and insect larvae dispose of the entire spectrum of strategies to survive flooding. Enchytraeids are simply too small for horizontal migration to non-flooded sites. The other groups lack physiological adaptations and seem to be less "prepared" for flooding. Table 4.2 lists survival times of several soil animal species in the field and in laboratory experiments; earthworms seem to endure inundation the longest.

4.2.1 Migration and recolonisation

Earthworms, as well as some terrestrial insect larvae (e.g., those of the Limoniidae) follow the ideal humidity at the flooding line and in banks of rivers and ponds (Topp 1981, Blanchart et al. 1987, Hentze-Diesing 1990, Ekschmitt 1991, Ausden et al. 2001). Also, terrestrial slugs retreat from floodwaters by active migration (Dahl et al. 1993). In a field experiment with completely flooded soil cores, hibernating invertebrates (adult Staphylinidae, Coleoptera larvae, and Araneae) emerged particularly during the first 10 days of flooding (Ausden et al. 2001). However, the distance covered by the smaller species will not be great enough to escape a fast-arriving inundation at a larger scale. Earthworms are the only soil invertebrates that are big enough to reach dryer habitats by active horizontal migration from a flooded site; single overland forays of 19 m have been recorded, even when the earthworms were not triggered by environmental stress (Mather and Christensen 1992). Consequently, an increased abundance is found in adjacent unflooded sites

(Ekschmitt 1991, Ausden et al. 2001, Tabeling and Düttmann 2002). For the smaller or slower soil invertebrates, vertical migration is a more successful way to escape from flood water. Gastropods climb onto floating vegetation, beneath bark, or into plants (Dahl et al. 1993). Diplopods (Zulka 1991), woodlice (Zerm 1999), and even earthworms (Pizl 1999) have been recorded to climb onto trees or other vertical emergent structures such as wooden poles in fences between pastures. Larvae of terrestrial Chironomidae have been observed to climb the vegetation layer (lichens on which they feed) during the rainy winter season when drought does not endanger them (Delettre 2000). However, it is not clear if the animals would use this possibility of vertical migration to seek refuge from flooded soil. Generally, earthworms (Brandsma 1997, 2002) and enchytraeids (Healy 1987, Beylich and Graefe 2002) concentrate in the upper centimeters of waterlogged soils, where oxygen supply is highest. In a field experiment, the earthworms *A. chlorotica, A. caliginosa, A. longa*, and *L. castaneus* moved upwards in soil samples (depth: 10 cm) when the lower halves were flooded; only 1 to 12.4% stayed in the flooded half compared to 42-54% in the lower halves of unflooded controls (Ausden et al. 2001). In certain soils, earthworms seek refuge in deeper mineral soil layers with sufficient air chambers ("drilosphere") during floods (Pizl 1999). Also Gastropods use air-filled cavities in the soil (Dahl et al. 1993). Some enchytraeids survive in the rhizophere (aerenchyma) in living and dead roots (Healy 1987).

Snails and slugs recolonize formerly flooded sites by passive transport: by the water itself, by driftwood, and by large animals (Dahl et al. 1993). Adult woodlice of the species *Trachelipus rathkii* resettle the habitat quickly from dryer refuges (Zulka 1989, Tajovský 1998). The flying adults of insects can cover great distances; thus, they recolonize formerly flooded sites rapidly (Hildebrandt 1995a).

4.2.2 Reproduction and flood-resistant stages

Cocoons are the most resistant development stages of earthworms, but whether juveniles or adults are more affected by flooding seems to vary with the species and with soil conditions (Hentze-Diesing 1990, Ekschmitt 1991, Meenken 1999). In a laboratory experiment, earthworm cocoons were hatched in oxygenated water and even young worms fed and grew in it (Roots 1956, Topp 1981). Especially, the common *L. rubellus* profits from its high reproduction rates. Out of the 90 cocoons ind.$^{-1}$ y^{-1}, about 15.8 individuals reach maturity (Evans and Guild 1948). Klok et al. (2004) reported that, in regularly inundated sites, adults of *L. rubellus* are smaller than in non-flooded sites. They reach maturity earlier and at a lower body weight, thus ensuring reproduction before the next flood. Also, the most common woodlice species in wetlands, *T. rathkii*, is an r-strategist that quickly builds up new populations. Its adults seem to be more tolerant towards flooding than juveniles (Zerm 1999). In contrast to this, the eggs and juveniles of chilopods endure winter and flooding, whereas the adults are very sensitive (Zerm 1999). Several chilopod (Zerm 1999) and earthworm species (*A. rosea, D. octaedra, D. rubidus, E. fetida, E. tetraedra, O. tyrtaeum, O. cyaneum, S. mammalis*; Sims and Gerard 1999) reproduce parthenogenetically, some enchytraeids are able to fragment (e.g. *C. glandulosa* and *C. sphagnetorum*; U. Graefe, pers. comm.; original observations by the author). This ability to reproduce without a sexual partner is an advantage in sites that are populated sparsely after inundation. In enchytraeids, embryogenesis and hatching from cocoons can be retarded under unfavorable conditions such as frost or drought (Giere and Hauschildt 1979) and is very likely to function as well during floods. For some terrestrial Dipteran larvae, the hatching temperature of hibernating eggs rises when they are flooded. As a consequence, mass hatching can be observed in spring when floodwaters flow back and temperatures rise (Hildebrandt 1995a).

4.2.3 Physiological adaptations

Earthworms and potworms profit from their cutaneous respiration that normally functions well under water at temperatures below 10°C. In spite of their thick cuticula and their open stigmata, some insect larvae as well are able to switch to cutaneous respiration to survive in water (Brauns 1954). These organisms are only affected by flooding if the oxygen content of the water diminishes and cutaneous respiration becomes difficult. The enchytraeid *Marionina filiformis* is a facultative anaerobic species (Beylich and Graefe 2002). There are several enchytraeid species, as well as insect larvae, with dense bristles that are used as a physical gill. They keep air bulbs on their body surface into which oxygen from the water diffuses, thus facilitating the cutaneous uptake of oxygen (Brauns 1954, Beylich and Graefe 2002). Earthworms have developed several physiological adaptations to overcome the lack of oxygen: *E. tetraedra* and *O. tyrtaeum* (Bouché 1977, Pizl 1999, Beylich and Graefe 2002) retain contact to the aerated zone by holding the tail vertically upwards and moving it within the water to maintain gas exchange. Also, aquatic

Oligochaetes (e.g., Tubificidae) that settle anaerobic sediments (Engelhardt 1986) use this "caudal respiration".

Respiratory pigments (haemoglobin) with a greater affinity for oxygen enable Lumbricids to take up oxygen from the water. This is already possible at an oxygen partial pressure of 19 mm Hg (atmosphere: 152 mm Hg). During anaerobic phases, earthworms synthesize and accumulate lactic acid and resynthesize glycogen when oxygen returns (Edwards and Lofty 1977, Lee 1985). Their calciferous glands add bicarbonate to the blood and bind CO_2 so it cannot hinder O_2 uptake (Graff 1983, Pizl 1999). Earthworms can change their osmoregulation, keeping the concentration of haemolymph constant. Water is excreted as quickly as it enters the body (Topp 1981). Diptera larvae achieve a certain resistance against swelling through reducing the concentration of body fluids (Hentze-Diesing 1990).

Earthworms can slow their metabolism when conditions become dry or cold (diapause stage, quiescence). Thus, winter and spring floods are better tolerated than summer floods (Topp 1981, Hentze-Diesing 1990). The ability to enter diapause is strongly developed in *A. chlorotica* (orig. observations) but weakly in *O. tyrtaeum*, which profits more from other physiological adaptations (a well-developed subcutaneous net of blood vessels, high concentrations of haemoglobin; Edwards and Lofty 1977). The formation of dense clusters (synaporia) of *Enchytraeus albidus* when flooded seems to be a general reaction to environmental stress and not only flood-specific (Ivleva 1969).

Table 4.2: Survival times of soil invertebrates under water

Taxa/ Author	Surviving Species	flooding duration	flooding circumstances
Earthworms (Ausden et al. 2001)	*Eiseniella tetraeda, Octolasion tyrtaeum, Allolobophora chlorotica* *Lumbricus rubellus*	270 d	September to May; site was irregularly summer-flooded
	Octolasion cyaneum	150 d	winter, field, alluvial gley
	Aporrectodea caliginosa, L. castaneus	120 d	laboratory experiment, in soil
	Dendrobaena octaedra, Satchellius mammalis	40 d	winter, field
Earthworms (Edwards and Lofty 1977)	*L. rubellus, A. chlorotica, A. rosea*	several weeks	in soil
Isopods (Zulka 1991)	*Ligidium hypnorum, Trachelipus rathkii, Hyloniscus riparius*	72 d	laboratory experiment
Diplopods (Zulka 1991)	*Polydesmus denticulatus*	22 d	laboratory experiment
Dipteran larvae (Priesner 1961)	*Tipula maxima*	some days	laboratory experiment; cold water

4.3 Soil invertebrates in flooded grassland: occurrence and effects of inundation

As inundations are a disturbing factor for terrestrial invertebrates, the following hypotheses were posed.

- There are no typical wetland species of soil macrofauna, only tolerant and well-adapted opportunists that become more abundant, while less frequent species disappear.

- Flooding of grassland reduces diversity, abundance, and biomass of all groups of soil macrofauna. The effects of winter flooding on soil fauna are less pronounced than those of summer flooding.

- In extensively flooded sites, soil animal groups with physiological adaptations have a greater diversity than those groups that only react by evasion when flooded.

4.3.1 Methods

To make sampling data of different authors available for a meta-analysis, all studied sites were classified in a gradient of inundation intensity from 0 to 6, taking regularity, duration, spatial dimension, and seasonal occurrence of flooding into account. The flooding intensities of the analyzed study sites were classified as follows:

0   non-flooded reference sites or sampling before introduction of flooding

1   episodic, irregular flooding, short (< 1 month), or at a small scale (highlands, ridges);

2   natural flooding in winter/spring about every 2 years;

3   regular flooding (natural or man-made) every winter/spring for 1-4 months;

4   like 3, > 4 months (extensively flooded);

5   summer flooding in the year of study;

6   fens and bogs, waterlogged (by re-wetting) most of the year.

Categories 5 and 6 were treated separately from the other categories since exceptional impacts of flooding on soil invertebrates during the warmer season can be assumed. Besides, the constant hydrologic conditions of fens and bogs (data only available for Lumbricids and Enchytraeids) are not comparable to the disturbing effect of flooding events. Species of the different groups were arranged in compacted tables according to their appearance in the most extensively flooded sites. The complete data tables are available on the internet (http://www.uft.uni-bremen.de/oekologie/nathalieplum2.htm). Data concerning abundance and biomass of whole groups were included in the tables, and means per category were calculated. When soil animals were sampled several times at one site, mean values of abundance and biomass were calculated for the table. Observations of different authors concerning immediate effects of flooding on population dynamics of the different soil fauna groups are compared in the following section. The quantitative sampling data (number of species, abundance, and biomass of soil macrofauna in flooded grassland of different inundation intensities) were used for a statistical meta-analysis.

4.3.2 Species composition of soil invertebrate groups in flooded grassland

Gastropoda. The most frequent snails were *Zonitoides nitidus, Carychium minimum, Succinea putris,* and *Cochlicopa lubrica* (Table 4.3). However, these species also occur in other biotopes. Slugs were not studied by most authors.

Table 4.3: Terrestrial Gastropoda in flooded grassland. The numbers indicate the number of analyzed sites of a certain inundation intensity in which the species occurs. Sources: Healy 1987, Dahl 1993, Spang 1996, Cejka 1999, Cejka 2004.

inundation intensity	6	4	3	2	1	total
	bog	> 4 months	1-4 months	every 2nd year	episodic	
number of sites	*1*	*8*	*2*	*3*	*2*	16
Carychium minimum (Müll.)	1		1	1		**3**
Deroceras laeve (Müll.)	1		1		1	**3**
Zonitoides nitidus (Müll.)			1			**1**
Succinea putris (L.)			1	2	1	**4**
Cochlicopa lubrica (Müll.)			2	3	1	**6**
Euconulus praticola (Reinhardt)			1			**1**
Succinella oblonga Drap.)			2	2	1	**5**
Nesovitrea hammonis (Ström)			1		1	**2**
Pseudotrichia rubiginosa (A.Schmidt)			1			**1**
Vallonia pulchella (Müll.)			2	3	2	**7**
Oxyloma elegans (Risso)						**0**
Arianta arbustorum (L.)				2		**2**
Carychium tridentatum (Risso)				2		**2**
Vitrea cristallina (Müll.)				2		**2**

inundation intensity	6	4	3	2	1	total
	bog	>4 months	1-4 months	every 2nd year	episodic	
number of sites	*1*	*8*	*2*	*3*	*2*	16
Monachoides incarnatus (Müll.)				2		**2**
Arion rufus (L.) agg.				2		**2**
Vertigo pygmaea (Drap.)				2		**2**
Cepaea hortensis (Müll.)				1		**1**
Cepaea nemoralis (L.)				1		**1**
Fruticicola fruticum (Müll.)				1		**1**
Trichia sericea (Drap.)				1		**1**
Eucobresia diaphana (Drap.)				1		**1**
Aegopinella nitidula (Drap.)				1		**1**
Vitrina pellucida (Müll.)				1		**1**
Monacha cartusiana (Müll.)				1		**1**

Lumbricidae. The most frequent species at the highest flooding intensities (3 - 5) were *Lumbricus rubellus, Aporrectodea caliginosa,* and *Eiseniella tetraedra,* followed by *Octolasion lacteum* (Table 4.4). This confirms Ekschmitt's (1991) contention that the earthworm coenosis he found in flooded marsh soils (eudominance of *L. rubellus* and *A. caliginosa* with *O. lacteum* and *E. tetraedra*) is the most resistant to flooding. The most constant species at all inundation intensities above 2 were *L. rubellus, A. caliginosa,* and *A. chlorotica*; three euryecious species that frequently occur in all types of habitats. The green morph of *A. chlorotica* var. *virescens* seems to be more abundant in wet soils, which can be related to the cryptic color compared to the conspicuous pink form. This is advantageous when the worms are forced to leave flooded burrows and become potentially subject to bird predation (Sims and Gerard 1985, Curry 1994, Ausden et al. 2001). In the indicator system of decomposer communities established by Beylich and Graefe (2002), *E. tetraedra, Octolasion tyrtaeum,* and *Helodrilus antipae* (which only occurred twice in the present studies) are indicators for wet soils. Pizl (1999) mentions *Octodrilus transpadanus* as another amphibian species. All other lumbricid species found in floodplain soils seem to be indifferent towards soil moisture conditions. In the analyzed studies, *O. tyrtaeum* occurred in only one of the extensively flooded sites, whereas *E. tetraedra* also occurred in unflooded sites and never dominated the earthworm fauna. Pizl (1999) observed that both rare subspecies of *E. tetraedra* (*E. tetraedra tetraedra* and *E. tetraedra intermedia*) prefer non-flooded habitats, while *E. tetraedra* has a tendency to prefer intensively inundated sites. All analyzed species occur in flooded and non-flooded sites; there are no typical floodplain species, only tolerant "hygrophilous" species, which corresponds with the conclusion of Spang (1996) that earthworms cannot be used as indicators of the "naturalness" of floodplain biotopes.

Table 4.4: Lumbricidae in flooded grassland. Sources: Healy 1987, Bolte and Moritz 1988, Ekschmitt 1991, Rodieck et al. 1992, Emmerling 1993, Makulec and Chmielewski 1994, Keplin et al. 1995, Spang 1996, Ma et al. 1997, Helling and Kämmerer 1998, Pizl and Tajovský 1998, Graefe 1998, Meenken 1999, Weigmann and Schumann 1999, Faber et al. 2000, Ausden et al. 2001, Beylich and Graefe 2002, Keplin and Broll 2002, Römbke et al. 2002, Tabeling and Düttmann 2002, Göbel 2003. The following subspecies were differentiated only by some authors and are therefore not considered as separate species in the analysis: *Allolobophora chlorotica virescens* (Emmerling 1993), *Aporrectodea caliginosa caliginosa, Aporrectodea caliginosa tuberculata* (Ma et al. 1997), *Aporrectodea caliginosa fa. trapezoides* (Bolte and Moritz 1988), *Dendrobaena rubidus tenuis*, and *Eiseniella tetraeda intermedia* (Pizl and Tajovský 1998), *Eiseniella tetraeda tetraeda* (Healy 1987, Pizl and Tajovský 1998), *Fitzgeringia platyura depressa* and *Helodrilus antipae tuberculatus* (Pizl and Tajovský 1998).

Table 4.4 Lumbricidae in flooded grassland (continued)

inundation intensity	6	5	4	3	2	1	0	to-tal
	bogs	summer flood	> 4 mths	1-4 mths	every 2nd year	epi-sodic	non-floo-ded	
number of sites	*13*	*2*	*13*	*18*	*20*	*21*	*5*	**86**
Lumbricus rubellus (L.)	10	1	10	14	18	19	5	**66**
Aporrectodea caliginosa (Sav.)	5	2	9	11	20	16	5	**59**
Eiseniella tetraeda (Sav.)	12	1	4	10	4	8	2	**35**
Octolasion lacteum (Örley)	8	2	3	7	4	3	3	**27**
Dendrobaena octaedra (Sav.)	9	2	3	5	3	5	1	**24**
Allolobophora chlorotica (Sav.)	5		2	5	12	7	2	**30**
Lumbricus castaneus (Sav.)	4		3	7	4	12	1	**26**
Octolasion cyaneum (Sav.)			2	1		8	1	**12**
Dendrodrilus rubidus (Sav.)			2		1			**3**
Octolasion tyrtaeum (Sav.)	2		1	3	4	1	1	**10**
Satchellius mammalis (Sav.)			1			3	1	**5**
Aporrectodea rosea (Sav.)	5	2		7	5	3	2	**19**
Lumbricus terrestris (L.)	3	2		4	10	5	1	**22**
Aporrectodea limicola (Michael.)	2			2	1			**3**
Helodrilus oculatus (Hoffmeister)	2			2				**2**
Aporrectodea longa (Ude)				1			1	**2**
Helodrilus antipae (Michael.)					2			**2**
Fitzgeringia platyura depressa (Rosa)					1	2		**3**
Allolobophora cupulifera (Tétry)						2		**2**
Lumbricus festivus (Sav.)							1	**1**

Enchytraeidae. Enchytraeids are a very diverse group that unfortunately has only been analyzed in a few winter-flooded sites (Graefe 1998), as well as in a fen (Healy 1987, Table 4.5). No study investigated extensively inundated sites; thus, it cannot be judged which species are the most common wetland species.

Table 4.5: Enchytraeidae in flooded grassland. Sources: Healy 1987, Beylich and Graefe 2002, Römbke 2002. Because of the scarce data, enchytraeid species are not sorted according to their appearance at the highest inundation intensities but simply according to their frequency of appearance in all categories.

inundation intensity	6	2	1	to-tal
	bog	every 2nd year	epi-sodic	
number of sites	*1*	*7*	*11*	*19*
Fridericia bulboides (Nielsen & Christensen)		6	11	17
Henlea perpusilla (Friend/Cernosvitov)	1	6	9	**16**
Enchytraeus buchholzi (Vejdovský)		5	9	**14**
Cognettia glandulosa (Michaelsen)	1	4	8	**13**
Henlea ventriculosa (d'Udekem)	1	3	9	**13**
Enchytraeus christenseni (Dózsa-Farkas)	1	4	7	**12**
Marionina argentea (Michaelsen)	1	3	4	**8**
Marionina vesiculata (Nielsen & Christensen)		3	5	**8**
Fridericia perrieri (Vejdovský)	1	1	4	**6**
Fridericia bisetosa (Lev.)		1	3	**4**
Fridericia connata (Bretscher)		2	1	**3**
Enchytraeus lacteus (Nielsen & Christensen)		1	1	**2**
Fridericia sylvatica (Healy)		2		**2**
Henlea nasuta (Eisen)		2		**2**
Enchytraeus coronatus (Nielsen & Christensen)		2		**2**
Marionina communis (Nielsen & Christensen)			4	**4**
Cernosvitoviella atrata (Bretscher)	1		2	**3**
Fridericia bulbosa (Rosa)			3	**3**
Marionina riparia (Bretscher)	1	1		**2**
Marionina filiformis (Nielsen & Christensen)		1	1	**2**
Fridericia alata (Nielsen & Christensen)			2	**2**
Fridericia ratzeli (Eisen sensu Nielsen & Christensen)			2	**2**
Achaeta unibulba (Graefe)			2	**2**

Table 4.5 Enchytraeidae in flooded grassland (continued)

inundation intensity	6	2	1	to-tal
	bog	every 2nd year	epi-sodi	
Buchholzia appendiculata (Buchholz)			2	**2**
Enchytronia minor (Möller)			2	**2**
Enchytronia parva (Nielsen & Christensen)			2	**2**
Cernosvitoviella goodhui (Healy)	1			**1**
Mesenchytraeus armatus (Levinsen)	1			**1**
Fridericia polychaeta (Bretscher)	1			**1**
Buchholzia fallax (Michaelsen)	1			**1**
Cognettia sphagnetorum (Vejdovský)		1		**1**
Fridericia galba (Hoffmeister)		1		**1**
Achaeta microcosmi (Heck and Römbke)		1		**1**
Marionina libra (Nielsen & Christensen)			1	**1**
Achaeta bibulba (Graefe)			1	**1**
Achaeta aberrans (Nielsen & Christensen)			1	**1**
Enchytraeus crypticus (Westheide & Graefe)			1	**1**
Fridericia maculata (Issel)			1	**1**

Isopoda. In total, isopods are rare in flooded grassland. The most frequent species is *Trachelipus rathkii;* all other species occur only occasionally or locally (Table 4.6). The most common species only occurred in unflooded reference sites (*Asellus vulgare, Porcellium scaber*) or were totally absent (*Oniscus asellus*).

Table 4.6: Isopoda in flooded grassland. Sources: Bolte and Moritz 1988, Handke et al. 1999, Emmerling 1993, Pizl and Tajovský 1998, Tajovský 1998, Zerm 1999, Beylich and Graefe 2002.

inundation intensity	5	4	3	2	1	0	total
	summer flood	> 4 mths	1-4 mths	every 2nd year	epi-sod.	non-floo-ded	
number of sites	*2*	*3*	*8*	*5*	*15*	*2*	**35**
Trachelipus rathkii (Brandt)	1		6	2	5	2	**15**
Trichoniscus pusillus (Brandt)	1			2	1	1	**4**
Haplophthalmus mengei (Zaddach)	1			1			**2**
Ligidium hypnorum (Cuv.)			1	1	3		**5**
Philoscia muscorum (Scopoli)			1		1		**2**
Hyloniscus riparius (Koch)				2		1	**3**
Porcellium conspersum (Koch)				1	1		**2**
Porcellio scaber (Latr.)						1	**1**
Armadillidium vulgare (Latr.)						1	**1**

Chilopoda and Diplopoda. Centipedes and diplopods are very sensitive to flooding and thus are generally rare in wetlands (Zerm 1999). Besides different *Lithobius*-species, *Lamyctes emarginatus* was the only chilopod species regularly found in flooded grassland (Table 4.7). Diplopods were practically absent in extensively flooded sites (Handke 1993, Zerm 1997, Handke et al. 1999, Weigmann and Wohlgemuth-von Reiche 1999). However, their diversity in moderately winter-flooded sites is remarkable (Table 4.8).

Table 4.7: Chilopoda in flooded grassland. Sources: Emmerling 1993, Tajovský 1998, Pizl and Tajovský 1998, Handke et al. 1999, Zerm 1999, Beylich and Graefe 2002, Voigtländer (unpublished data)

inundation intensity	5	4	3	2	1	0	total
	sum-mer flood	> 4 mths	1-4 mths	every 2nd year	epi-sod.	non-floo-ded	
number of sites	*2*	*3*	*6*	*5*	*19*	*1*	*36*
Lamyctes emarginatus (Newport)		3	6		2	1	*12*
Lithobius curtipes (C.L. Koch)			2		2		*4*
Lithobius forficatus (L.)			1			1	*2*
Lithobius microps (Mein.)			1			1	*2*
Strigamia acuminata (Leach)					1		*1*

Table 4.8: Diplopoda in flooded grassland. Sources: Emmerling 1993, Tajovský 1998, Pizl and Tajovský 1998, Handke et al. 1999, Zerm 1999, Beylich and Graefe 2002, Voigtländer (unpublished data)

inundation intensity	5	4	3	2	1	0	total
	sum-mer flood	> 4 mths	1-4 mths	every 2nd year	epi-sodic	non-floo-ded	
number of sites	2	3	6	5	19	1	36
Polydesmus inconstans (Latz.)			2		3	1	6
Leptoiulus cibdellus (Chamberlin)			2				2
Julus scandinavius (Latz.)			1		2		3
Craspedosoma rawlinsii (Leach)			1			1	2
Cylindroiulus caeruleocinctus (Wood)					7		7
Cylindroiulus nitidus (Verh.)					6		6
Polydesmus denticulatus (C.L. Koch)					5		5
Brachyiulus pusillus (Leach)					3		3

Table 4.8 Diplopoda in flooded grassland (continued)

	3	1	4
Brachydesmus superus (Latz.)	3	1	4
Leptoiulus belgicus (Latz.)	3		3
Choneiulus palmatus (Nemec)	1		1
Craspedosoma alemannicum (Verh.)	1		1
Tachypodoiulus niger (Leach)	1		1
Xestoiulus laeticollis (Por.)	1		1
Leptoiulus proximus (Nemec)	1		1
Glomeris tetrasticha (Brandt)	1		1
Ommatoiulus sabulosus (L.)		1	1
Proteroiulus fuscus (Am		1	1

Insect Larvae. Data on terrestrial insect larvae in wetlands are scarce. Because of the wide size ranges and determination difficulties, most authors only studied selected families or genera. Species were not always determined. For this reason, a statistical meta-analysis was not possible. Large, soil-dwelling insect larvae such as those of Tipulidae (crane flies/leatherjackets), Bibionidae (March flies), and Elateridae (click beetles; Ausden et al. 2001), Staphylinidae, and Carabidae (Rusek 1984, Hansen and Castelle 1999) are frequently sampled together with earthworms in macrofauna studies, where families with smaller larvae are not recorded. In different peat meadows, the Chironomidae were the dominating dipteran family next to Limoniidae, Ceratopogonidae, Sciaridae, Phoridae, Chloropidae, and Muscidae (Frouz and Syrovátka 1995). For further lists of occurring families, see the complete table on the internet (http://www.uft.uni-bremen.de/oekologie/nathalieplum2.htm).

4.3.3 Immediate effects of flooding on soil invertebrates

Table 4.9 shows the observed immediate effects of flooding on the studied soil invertebrate groups. In earthworms (the best studied group), all effects were observed. It is possible that these effects also occur in the other groups but have not been observed in the few studies available.

Extensive flooding caused species of all groups to disappear (terrestrial gastropods: Borcherding 1889, Schröder 1980, Dahl et al. 1993, Handke 1993, Spang 1996, Cejka 1999; earthworms: Ekschmitt 1991, Faber et al. 2000, Ausden et al. 2001;

isopods: Zerm 1997, Pizl and Tajovský 1998, Handke et al. 1999; diplopods: Zerm 1997, Handke et al. 1999; chilopods: Handke et al. 1999). In sites flooded by polluted river water, some species of the expected earthworm and enchytraeid coenoses were absent and abundances were lower than expected (Römbke et al. 2002). Especially, anecic (i.e., deep-burrowing) earthworms disappear in inundated sites (Keplin et al. 1995, Graefe 1998). They obviously tolerate only low inundation intensities, probably because waterlogged soils or high ground-water tables impede their burrowing and thus limit their habitat. Only in one case, *L. terrestris* was still present even after a disastrous summer flood (Pizl and Tajovský 1998). Its general presence in this study (as well as the high number of 7 earthworm species, Table 4.4) can be explained by the normally low inundation intensity of the site (regular spring floods and irregular autumn floods from several days up to two weeks; K. Tajovský, pers. comm.). Besides, a quick recolonisation of the site from refuges by this large and mobile species can be assumed.

Also, the abundance of nearly all groups was reduced by extensive flooding (earthworms: Volz 1976, Ekschmitt 1991, Graefe 1998, Faber et al. 2000, Römbke et al. 2002, Tabeling and Düttmann 2002; chilopods: Zerm 1997, Pizl 1998, Handke et al. 1999). For isopods, diplopods, and chilopods, the effects became stronger with increasing inundation duration (Zerm 1997, Pizl 1998). In a field experiment, Dipterans of the genus *Bibio* (Geoffr.) emerged only from unflooded soils cores (195 ind. m^{-2}) but not from formerly flooded soil samples (Ausden et al. 2001). Earthworm populations were locally reduced by 80 to 100 % (Ekschmitt 1991, Faber et al. 2000), while in adjacent non-flooded sites, they increased up to 500%, partially due to migrating earthworms (Ekschmitt 1991). In winter-flooded and summer-wet sites, an untypical maximum of earthworm populations in summer with a decline in autumn could be observed (Rodieck et al. 1992). Chilopods can recover from resistant stages and occur in large numbers in spring after winter floods (Zerm 1999).

In all studies where earthworm biomass (total and individual) was determined, this was reduced in flooded sites (Keplin et al. 1995, Graefe 1998, Meenken 1999, Faber et al. 2000, Ausden et al. 2001, Göbel 2003). Ekschmitt (1991) found a correlation between winter flooding intensity and the reduction of total as well as individual biomass in earthworms and Tipulid larvae. This can be explained by the reduction of abundance and by reduced food uptake during diapause and higher respiration rates while fleeing (Ekschmitt 1991). Pizl (1999) considered the impact of episodic summer floods on the earthworm communities of a hardwood forest to be much more severe than that of regular winter flooding. The population of chilopods collapsed during a summer flood (Pizl 1998).

Nevertheless, there are some exceptions to this general pattern. In highland brooks, an increase of the total soil fauna diversity after short and spatially limited flooding was observed (Emmerling 1993). Cejka (2004) observed more gastropod species and greater abundances at low or medium inundation intensities compared to drier sites (Table 4.3, Figure 4.1a). New species of polyhygrophilous (amphibious) and freshwater gastropods appeared (Dahl et al. 1993, Obrdlik et al. 1995, Spang 1996, Cejka 2004), just as aquatic Dipteran families (Hildebrandt 1995b, Hansen and Castelle 1999). Thus, diversity in total, including terrestrial as well as aquatic species, is rather raised by flooding. In one case, Dipteran abundance was increased after flooding (Rusek 1984). Ekschmitt (1991) observed that controlled flooding can raise earthworm and dipteran abundance in years with dry summers. Re-wetting of fens and bogs can have a positive influence on the occurrence of soil invertebrates: New earthworm species (*Helodrilus oculata* and *Aporrectodea limicola*) appeared, earthworm abundance increased locally (Keplin et al. 1995), and Dipteran diversity was also raised (Frouz and Syrovátka 1995, Frouz 2000). The micro-relief of typical bogs with hummocks as potential refuges may play a role within this scope. In a quaking marsh (category 6), enchytraeid diversity was very high, but abundances low. Maximum values were reached when the surface was flooded in April, compared to a dry July, when the water level fell to 2 cm beneath the soil surface (Healy 1987).

4.3.4 Soil invertebrate populations at different lengths of inundation

The influence of the general site factor "inundation intensity" on mean species number, abundance, and biomass of each animal group was tested using sampling data of different authors. When the statistical pre-conditions were met, One-Way-ANOVAs were calculated (followed by Bonferroni tests for pair comparisons); otherwise a Kruskal-Wallis' H-Test (followed by the Tamhane test) was used.

Number of Species. A significant influence of inundation intensity on species number was found for earthworms ($F = 2.3$, $p < 0.05$, df = 6; Figure 4.1b), chilopods ($H = 13.8$, $p < 0.05$, df = 5; Table 4.7), and diplopods ($H = 12.0$, $p < 0.05$, df = 5; Table 4.8, Figure 4.1f). For enchytraeids and isopods, only a trend can be seen (H-Test, n.s., Figures 4.1c, e). In sites of inundation intensity 4 (winter flooding for more than 4 months), the fewest earthworm and chilopod species and no isopods nor diplopods were found (Figure 4.1b, e, f). Fens and summer flooded sites have earthworm and enchytraeid species numbers in the order of the non-flooded reference sites, probably because of the rather stable hydrologic conditions (Figures 4.1b, c). Natural floodplains flooded every second winter (category 2) showed exceptionally large Gastropod species numbers (Table 4.3, Figure 4.1a). For this group, the agricultural land use of the sites was also tested, taking fallow land and pastures as the lowest and the number of mowing events per year as an increasing stress factor. The intercept effect of both parameters, inundation and agriculture, on Gastropod species numbers is significant (multivariate test, Wilks-Lambda, $p < 0.05$).

Abundance and Biomass. For abundance and biomass, parameters that depend strongly on seasonality, less data were available. However, there is a trend toward lower isopod abundances with increasing inundation intensity (H-Test, n.s). Figure 4.1b shows that the extensively flooded sites together with the summer flooded sites, have the lowest earthworm abundances, distinctly lower than fens and bogs. Thus, it can be assumed that summer floods have a larger but transitory impact on earthworm abundance - namely at the same scale as regular intensive winter flooding - than on species number.

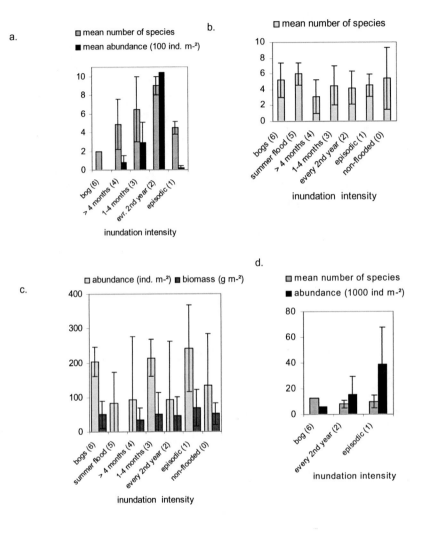

Figure 4.1: Soil invertebrate populations at different inundation intensities. a. Gastropod species number and abundance; b. earthworm species number; c. earthworm abundance and biomass; d. enchytraeid species number and abundance; Mean values (with standard deviations) from sampling data of different authors. Sources: see Tables 4.4 to 4.9.

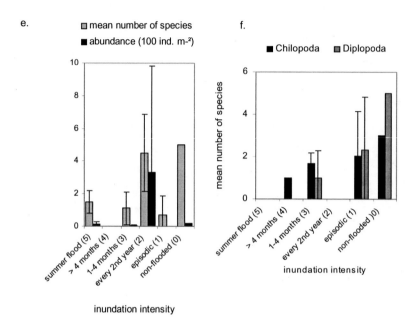

Figure 4.1 (continued): Soil invertebrate populations at different inundation intensities. e. isopod species number; f. chilopod and diplopod species number. Mean values (with standard deviations) from sampling data of different authors. Sources: see Tables 4.4 to 4.9.

Table 4.9: Effects of flooding on different soil animal groups. Sources: Volz 1976, Schröder 1980, Rusek 1984, Hentze-Diesing 1990, Ekschmitt 1991, Emmerling 1993, Handke 1993, Keplin et al. 1995, Spang 1996, Graefe 1998, Helling and Kämmerer 1998, Tajovský 1998, Cejka 1999, Handke et al. 1999, Hansen and Castelle 1999, Meenken 1999, Weigmann and Wohlgemuth-von Reiche 1999, Zerm 1999, Faber et al. 2000, Frouz 2000, Ausden et al. 2001, Keplin and Broll 2002, Römbke et al. 2002, Tabeling and Düttmann 2002, Cejka 2004.

Number of studies	3	14	5	6	6
Effect	Gastro-pods	Lumbri-cids	Iso-pods	Diplopods Chilopods	Dipteran larvae
Disappearance of species	x	x	x	x	x
Change in species composition	x	x			x
Appearance of new species	x	x			x
Increase in diversity		x	x	x	x
Reduced total abundance	x	x	x	x	
Total breakdown of population	x	x	x	x	
Increase in abundance	x	x	x	x	x
Reduced total biomass		x			x
Reduced individual biomass		x			x
Impact on population structure		x	x		

4.3.5 Methodological remarks on the meta-analysis

In comparing different soil-fauna studies, problems occur due to different sampling methods and time scales. Some data used for the meta-analysis are single values, some are means of several samples taken through time. A further problem is the missing specification of the sampling design, abiotic conditions (season, climatic conditions, flooding regime), sampling sites, and/or community structure. Species composition is the most reliable parameter for the animal groups with a small database. Quantitative data on soil animals strongly depend on the weather conditions and fluctuate in the course of a year, also independently from hydrologic conditions. None of the studies recorded all soil animal groups on one site. However, trends can be seen and useful recommendations for wetland management can be derived.

4.4 The role of soil invertebrates in flooded grassland ecosystems

4.4.1 Influence on soil properties

The influence of soil animals on physical and chemical soil properties has not yet been studied in floodplain ecosystems. However, the contribution of earthworms to the recovery of compacted, consolidated, and unstructured soils after floods can be assumed: by their burrowing activity, endogeic (soil-dwelling) earthworms such as *A. caliginosa* and *A. chlorotica* re-establish soil structure (pores blocked by alluvial deposits) and improve the conditions for plant root growth in heavy soils that are widespread in floodplain marshes (Ausden et al. 2001). Their burrowing also increased infiltration capacity after heavy rains (Sharpley et al. 1979).

4.4.2 Invertebrates as prey

Conservation of wetlands and water management often aim at protecting habitats for birds, especially waders, which partially feed on terrestrial invertebrates. The significance of snails and slugs as prey is probably high, but there are no data available. Earthworms in general, as well as big insect larvae like those of the Tipulidae, are a valuable prey for wading birds because they are easy to catch and their biomass is large (Brodmann and Reyer 1999). The common snipe (*Gallinago gallinago* L.) and the black-headed gull (*Larus ridibundus* L.) feed exclusively on earthworms. Young lapwings (*Vanellus vanellus* L.), blacktailed godwit (*Limosa limosa* L.), redshank (*Tringa totanus* L.), and oystercatcher (*Haematopus ostralegus* L.) probe regularly for them (Faida et al. 2003). In a field experiment with central cages excluding birds, earthworm biomass was reduced by up to 40% by bird predation (Ekschmitt 1991). In soft, waterlogged patches of a marsh soil, the large earthworm abundance in early spring was even reduced to zero during the breeding season (Ausden et al. 2001). The influence of birds on dipteran larvae populations is difficult to determine because of their heterogeneous distribution in soil (Ekschmitt 1991).

Tipulid larvae reach their maximum body weight during the breeding season (Priesner 1961) and, in years of mass occurrence, can surpass earthworm biomass (Brandsma 1997). However, it seems that some birds avoid them, probably because of their bad taste or the toxins they contain (Brodmann and Reyer 1999). Especially young lapwings and blacktailed godwits also feed on aquatic dipteran larvae, emerged adults, and even large enchytraeids (A. Schoppenhorst, pers. comm.). The stomach analysis of fish in wetlands showed that earthworms can account for a large part of their food as well (Reimer and Zulka 1994).

Not only the presence, but also the accessibility of prey, guarantees the breeding success of wading birds. Available soil invertebrate prey is defined as the biomass of macrofauna in the upper layer of soil that is soft enough to probe in (Ekschmitt 1991, Brandsma 1997). After flooding, the available prey in general (total biomass) and/or the profitability of the single soil invertebrate as prey (individual biomass) is reduced. In waterlogged or flooded soils, most soil animals are found in the upper layer. There, they are available to all probing birds, also to those with short beaks that can only reach depths of less than 3 cm (such as lapwings; Ausden et al. 2001). Especially, earthworms vacating their flooded burrows and moving aboveground or concentrating in non-flooded patches (food patches, Tabeling and Düttmann 2002) become easy prey for wading birds. Additionally, a waterlogged soil has a lower penetration resistance and allows more bird species to probe for their prey. Threshold values differ for different bird species (Faida et al. 2003). For example, common snipes leave their breeding areas when penetration resistance of the soil rises above 6 kg (Düttmann and Emmerling 2001). Ekschmitt (1991) mentioned a maximum "penetrability" of 125 N m^{-2} for lapwings. At higher values, which are reached in late spring when the soil is drying, birds do not succeed in probing and have to switch to other sites or food resources. Earthworm populations can then develop rapidly, until the soil also becomes too hard for them (Göbel 2003). A macrofauna fresh weight biomass below 10 g m^{-2} led to a severe decline of breeding blacktailed godwits in a wet meadow (Brandsma 1997). Bird populations can only recover at a minimal macrofauna biomass of 25-30 g m^{-2}. An earthworm biomass of 45 g m^{-2} was found to be sufficient even for large densities of breeding lapwings and blacktailed godwits (Brandsma 1997, 2002).

4.5 General discussion

The results collected in this study show that among terrestrial invertebrates in flooded temperate grassland, there are no species restricted to wetlands, only tolerant, hygrophilous species and well-adapted generalists and opportunists that are favored by flooding. Less frequent or weakly adapted species disappear with extensive inundation. As observed by many different authors, abundances and biomass of soil invertebrates are immediately reduced by flooding. The effect becomes stronger with flooding duration and rising temperatures, but even then, it is reversible and will normally be compensated to a large extent during the next soil-dry period. Only short and small-scaled flooding, as well as wetting of fens and bogs, can attract new soil fauna species and raise abundances. Dipterans, but also chilopods, may occur in greater abundances after flooding.

In sites that are flooded at least every second year, the groups without physiological adaptations to survive inundations are represented by very few species (Isopods: 7, Chilopods: 5, Diplopods: 3 species, respectively). Gastropods are an exception; although they only react to inundation with evasion and recolonisation, they have the same diversity in regularly flooded sites as the well-adapted enchytraeids (25 species), followed by earthworms (18 species). The diversity of insect larvae is probably the greatest by far, as the number of recorded Dipteran families is already 38. Earthworms and enchytraeids, as well as insect larvae, generally prefer wet conditions, and they profit from their physiological adaptations for survival in sufficiently aerated water. However, when oxygen in stagnating flood water is reduced to a level below their tolerance, even the physiologically adapted groups are affected by flooding. A general trend towards reduced species numbers and lower abundances with increasing inundation intensity was not only found for isopods and millipedes but also for earthworms. This is probably due to their low reproduction rate and slow generation cycle compared to enchytraeids and insects. Here again, Gastropods are an exception, as they occur in greater species numbers in sites with regular, moderate winter flooding. Fens and bogs, due to the rather constant hydrologic conditions, cannot be compared with flooded sites. They are settled by more earthworm and enchytraeid species than extensively flooded sites. The effect of episodic summer floods is less pronounced than that of regular extensive winter inundations. As far as research has been done, earthworms seem to play the most important role in wetlands as prey for birds and by influencing soil physical properties. Thus, the following recommendations for water management focus on this group.

4.6 Recommendations for wetland management and research needs

Especially where backwater flooding is practiced to promote wetland birds, soil fauna as their food resource should be spared or promoted by water management. Naturally, a rising flood comes rather quickly. However, a slow rising of water levels in wetlands (which can often be achieved with controlled flooding) can be favorable for soil animals, giving them the chance to seek refuge or to produce eggs and cocoons. The retreat of the water, naturally some decimeters per day, should be slow as well. The "moving littoral" reveals food for land animals and gives aquatic organisms the chance to move with the water (Dohle et al. 1999). Isopods, chilopods, land snails, and earthworms make use of non-flooded habitats as refuges; most of all small woods, reeds, and fallow meadows should be conserved to facilitate the recolonisation of grassland. Several short inundations seem to be less severe for soil animals than a single extensive flood (Beylich and Graefe 2002). Short winter flooding can promote soil fauna, especially earthworms, while extensive flooding can considerably reduce their spring abundances and biomass. Instead of backwater flooding, it would be better to keep most of the soil waterlogged and soft and to let it fall dry enough for earthworm populations to build up in early spring (Keplin et al. 1995, Schekkermann 1997, Ausden et al. 2001). Definitely, a mosaic of different habitats containing open water surfaces, shallow pools, vegetation-free sites, flooded and unflooded grassland, as well as forest patches, will be favorable for both birds and invertebrates (Ekschmitt 1991, Ausden et al. 2001, W. Eikhorst, pers. comm.). A micro-relief structure with shallow ditches and drier ridges as is typical for meadows in Northern Germany can partly fulfill this claim.

Some observations indicate that, when flooding is (re-)introduced in grassland, it attracts large numbers of wading birds but that this attractiveness weakens after some years. This is probably due to the negative effect of regular flooding on the spring populations of soil macrofauna. Pausing of flooding for one or two winters in 10 years could be a possibility to allow earthworm and other soil invertebrate prey populations to recover (A. Schoppenhorst, pers. comm.). However, measures such as fertilizing and liming of wet meadows in order to raise soil fauna biomass and promote breeding birds, as proposed by Brandsma (2002), should take other aspects concerned with this into account, such as natural characteristics of the soil (especially of oligotrophic peat with a naturally low pH), vegetation, and the quality of the water related to the sites. When (formerly) fertilized grassland is flooded, the vegetation should be removed by grazing or mowing prior to flooding to prevent anoxic conditions harmful to the soil fauna, or water levels should be allowed to fluctuate so that oxygen supply is replenished during periods of drying out (Ausden et al. 2001).

As this review shows, quite a number of studies on earthworms in wetlands have been carried out, focusing on their role as prey for wetland birds. Little is known about other soil macrofauna groups and enchytraeids. Research concerning the role of soil fauna in floodplain ecosystem processes (infiltration, decomposition, mobilisation of substances, etc.) with different water management is needed.

4.7 Acknowledgements

I thank Juliane Filser, as well as two anonymous referees, for reviewing the manuscript. I am also grateful to David Russell and Karin Nitsch for linguistic correction, Matty Berg, Karsten Schröder, and Karin Voigtländer for the taxonomic revision of the species lists and several of the cited authors (Malcolm Ausden, Heinrich Belting, Obe Brandsma, Tomas Cejka, Klemens Ekschmitt, Ulfert Graefe, Wei-Chun Ma, Heinrich Tabeling, Karel Tajovský) for additional information about their studies.

5. Floods and drought: response of earthworms and potworms (Oligochaeta: Lumbricidae, Enchytraeidae) to hydrological extremes in wet grassland

published as: Plum, N.M. and Filser, J. (2005): Floods and drought: response of earthworms and potworms (Oligochaeta: Lumbricidae, Enchytraeidae) to hydrological extremes in wet grassland. *Pedobiologia* 49 (3) 443-453

Abstract: The population dynamics of earthworms and potworms (Oligochaeta: Lumbricidae, Enchytraeidae) were recorded in three floodplain meadows differing in soil type in Northern Germany during two years of contrasting hydrological extremes: 2002 with high precipitation, extended winter inundations and summer floods in two of the sites, and 2003 with an extraordinarily dry summer. In all sites, earthworms dominated the biomass of soil fauna, the dominant species being *Octolasion tyrtaeum* in marsh soil, *Octolasion cyaneum* in peat soil, *Allolobophora chlorotica* in gley soil, and *Lumbricus rubellus* in all three sites. The enchytraeid communities, composed of 7-10 species and differed widely between sites.

While the summer flood of 2002 reduced annelid populations in peat soil to zero, in gley soil only enchytraeids were sharply reduced. The earthworms there probably were less affected because of shorter duration of the flood, the lower air temperatures and SOM contents and the diapause of the dominant *A. chlorotica*. During summer drought in 2003, earthworms and enchytraeids were absent from gley soil, while they had maximum abundances in peat soil. The marsh soil, structured by a microrelief of shallow ditches and ridges, had highest densities of annelids. As it was not subjected to summer flooding in 2002, populations of both earthworms and potworms were stable. The drought in 2003 led to a sharp decrease in earthworm abundance in July, while the enchytraeids reached maximum values. Dynamics and survival strategies of different annelid species are discussed.

The water holding capacity as well as the organic matter content of the soil are important factor for annelids in sites subjected to such hydrological extremes. To promote terrestrial annelids, controlled flooding should be kept short, especially in winters following natural summer floods. If possible, recovery times for annelids should be guarenteed, that is 2-3 months without inundations for enchytraeids and about half a year for earthworms. A microrelief with drier refuge sites can be judged as favourable for earthworms. An artificially raised groundwater table will favour annelids during drought periods.

Keywords: invertebrates; Lumbricidae; Enchytraeidae; population dynamics; wetland; inundation; drought

5.1 Introduction

Under natural conditions, floodplain meadows in Europe are irregularly flooded during winter. To attract water birds and waders or to promote special plant communities, controlled flooding is often brought about every winter and extended to the spring months (Haslam 2003, G. Oertel and B. Olbrich, pers. comm.). Soil fauna as an important part of the ecosystem, in wetlands often dominated by annelid worms, may suffer from too extensive flooding, especially when additional hydrological stress situations induced by summer floods or drought occur.

Earthworm abundance and biomass usually is reduced by extensive flooding at a large spatial scale (Keplin et al. 1995, Graefe 1998, Ausden et al. 2001, Römbke et al. 2002), locally by 80 to 100 % (K. Ekschmitt, pers. comm., Faber et al. 2000). However, in years with dry summers, controlled flooding can raise earthworm abundance (K. Ekschmitt, pers. comm.). A statistical meta-analysis of 86 earthworm inventories showed that on extensively winter- and spring-inundated and summer-flooded sites, less earthworm species occur and abundances are lower (Chapter 4). Little is known about enchytraeids in flooded sites. In regularly winter-flooded sites (Graefe 1998) as well as in bogs (Healy 1987) they occur as a very diverse group, but compared to well-drained soils their density is rather low in very wet sites (Beylich and Graefe 2002).

In the present study, the population dynamics of earthworms and potworms were recorded during two exceptional years: the wet year 2002 with flood events during the warmer season in two of the three study sites, followed by a very dry and hot summer in 2003. It was hypothesized that (i) the species composition of earthworms and enchytraeids and their overall density differs according to site conditions; (ii) inundations reduces abundance and biomass of annelids worms, except for some well-adapted species; (iii) summer inundation and (iv) summer drought has a greater impact on annelid populations than regular winter inundations, and (v) enchytraeids are better adapted to inundations than earthworms and recover more rapidly from such events.

5.2 Material and methods

5.2.1 Study sites

Sampling of annelids was carried out in three regularly winter-flooded meadows along rivers in Northern Germany differing in soil type (river Wümme: peat, river Ochtum: marsh soil; river Elbe: contaminated gley soil; for site characteristics see Table 5.1). The peat and the marsh sites are situated in nature reserves; the meadows are mown once or twice a year. Both sites are surrounded by ditches which convey the flood water. The marsh site is structured by a man-made micro-relief consisting of shallow ditches, ensuring the drainage of the ridges in-between even under very wet conditions. However, in winter the ridges also are flooded by water management from December to February (B. Olbrich, pers. comm.). The peat site is also subject to hydrological management, although at high water levels flooding happens naturally almost every winter, and episodic summer floods occur. The gley site, a meadow that is mown twice or three times a year, but not fertilised, is subjected to various contaminants from the river water (soil values for dioxins: 280 mg kg^{-1}, Pb: 320 mg kg^{-1}, Cu: 360 mg kg^{-1}, locally also PAH and furanes; Kleefisch and Knes 1997).

5.2.2 Weather and hydrological dynamics

The sum of precipitation in Bremen in the year 2002 broke all records since the beginning of meteorological survey. With 2000 mm, it exceeded the long-term mean by 150% (Deutscher Wetterdienst 2003). Heavy summer rains led to repeated short-term accumulation of water in the shallow ditches of the study site at the river Ochtum (marsh soil, July 14 to August 2). As this river has a very small catchment, rainfall did not result in a summer flood on this site. Thus, it can be used as non- flooded reference, offering dry refuges for annelids on the ridges even in the wettest phases of the summer of 2002.

Table 5.1: Vegetation and soil characteristics of the study sites

Main data sources: [a] Erber 1998; [b] B. Olbrich, pers. comm.; [c] B. Schuster, unpublished data; [d] Janhoff 1992; [e] G. Oertel, pers. comm., [f] Kleefisch and Knes 1997; [g] own measurements / observations; [h] www.wetteronline.de/reuro.htm, 21.10.2004.

Study site	Nature reserve "Ochtumpolder bei Brokhuch-ting", Bremen	Nature reserve "Borgfelder Wümmewiesen, Bremen	Soil monitoring site of the State Lower Saxony, Gorleben/Elbe
Soil type	marsh soil[a]	peat (10-15 cm) on river sand[d]	river gley soil[f]
pH ($CaCl_2$)	4.8 [a]	4.6 [c, d]	5.5[f]
N_t (%)	0.7[a]	2.3 [c, d]	0.7[f]
C_{org} (%)	6.2[a]	29.1 [d]	9.5 [f]
C/N	9.1[a]	14.8 [c, d]	13.4[f]
SOM (%)	> 10[a]	27.8 [c, d]	9.3 [f]
Field capacity (vol.-%)	64[a]	67 [d]	64[f]
Water holding capacity (mass %)	64[a]	76 [g]	52[f]
Special characteristics	oxide concretions[a]	peat mixed with river sediments[d]	contamination (see text) [f]
River (tributary to stream)	Ochtum (Weser)	Wümme (Weser)	Elbe
Climate	oceanic	oceanic	subcontinental
Water management, inundation interval	backwater flooding every winter from December to February[b]	backwater flooding in winter, episodic summer floods at high water levels [d]	no management, natural inundations every second winter, episodic summer floods[f]
Inundations before and during the study period (exact dates only when observed by the authors)	Dec. 01 to Feb. 02, (rainwater 07-24 to 08-02-2002), 12-15-02 to 02-15-03 [g]	Sept. 01, Nov. 01 to Apr 02, 07-20 to 08-21-2002, 10-22-02 to 03-15-03 [g]	Nov. 01 to Feb. 02, 08-18 to 09-11-2002, Nov 02 to Feb. 03 [g]
Plant association	Phalaridetum arundinaceae [a, g]	Caricetum gracilis (Calthion) [d, g]	Phalaridetum arundinaceae, Glycerietum fluitantis [f, g]

Table 5.1 (continued): Vegetation and soil characteristics of the study sites

Study site	Nature reserve "Ochtumpolder bei Brokhuch-ting", Bremen	Nature reserve "Borgfelder Wümmewiesen, Bremen	Soil monitoring site of the State Lower Saxony, Gorleben/Elbe
Root horizon	10 cm [a]	13 cm [d]	7 cm [f]
Agricultural use	mowing 1-2 times/year; sometimes pasture in autumn (not in years of study) [b]	mowing 1-2 times/year [e]	mowing 2-3 times/year [f]
precipita-tion (mm y⁻¹) 2002	1070 [h]	1070 [h]	852 [h]
2003	607 [h]	607 [h]	427 [h]
sun hours 2002	1523 [h]	1523 [h]	1501 [h]
sun hours 2003	1885 [h]	1885 [h]	2005 [h]

In contrast, the nature reserve „Borgfelder Wümmewiesen" (peat soil) had been subjected to an autumn flood in September 2001, an exceptionally extended winter flood until late April 2002, followed by a summer flood in 2002 (July 20 to August 21 at air temperatures of 20-30°C) and an early winter-flood after a heavy rainstorm on October 22. Hydrological conditions normalised only in spring 2003 when the water was allowed to leave the meadows in early March and the exceptional dry spring and summer of 2003 subjected the region to another hydrological extreme.

The Gorleben site (gley soil) is situated in the east of Lower Saxony at the middle course of the river Elbe which statistically overflows its banks every second winter (B. Kleefisch, pers. comm.). In fact, the site was inundated from early November to late February in both winters 2001/2002 and 2002/2003. Heavy rains on the remote catchments in the first two weeks of August 2002 caused an enormous flood that reached the Gorleben site downstream on August 18 and stayed until September 11 at moderate temperatures of 13-22°C. Compared to Bremen, the Gorleben region recorded a lower annual precipitation in both years of study and more sun hours in 2003. Thus, the summer drought of 2003 at this site was even more marked (2005 sun hours and 427 mm precipitation compared to 1885 sun hours and 607 mm in Bremen, Deutscher Wetterdienst 2004, www.wetteronline.de/reuro.htm, 21.10.2004).

5.2.3 Sampling methods and statistical analysis

Sampling of annelids was carried out 10 times between June 2002 and September 2003 at irregular intervals depending on hydrological events (before and after floods, in a period of drought). At each date and site, the actual soil water content (in the upper 4 cm) and enchytraeids were sampled in 5 randomly distributed plots, avoiding areas with traces of preceding samplings. Earthworms were only sampled if field conditions permitted it, i.e. when the site was free of stagnant water (see Fig. 5.1 and 5.2). In the marsh site with the microrelief, samples were always taken at the "slope" of the shallow ditches.

As two of the study sites are situated in nature reserves, the extraction of earthworms was not done with formalin (ISO 2004), but with a suspension of hot mustard (Högger 1993). One hundred g of ground green mustard seed (Develey, Germany) were allowed to swell for at least 2 h in 1 l of water. Immediately before application this suspension was diluted to 3 x 10 l in plastic canisters, giving a concentration of 3.3 g l^{-1}. To prepare the extraction site, vegetation was cut short and removed plant material was searched for earthworms. The suspension was distributed evenly over the ground inside a wooden frame (50 x 50 cm) until the soil was waterlogged, and a second time after 10 minutes (maximum: 10 l of suspension). Earthworms as well as all other macroinvertebrates were collected in a wetted plastic bag until 30 min. after the first application. As a control, a soil sample of 25 x 25 x 20 cm (a depth always reaching the groundwater table of these sites) on the same surface was dug with a spade and hand-sorted. Earthworms were determined according to Sims and Gerard (1985). The biomass of living adults and juveniles was estimated by a volumetric method (displacement of water in a scaled vessel, 1 ml being equivalent to 1.064 g fresh biomass; Dunger and Fiedler 1989). Soil samples (Ø = 8 cm, 12 cm deep, 5 replicates) were taken with a soil corer and transported to the laboratory in plastic bags. This way of sampling was also possible during floods. The core was halved vertically and one half was used for cold wet funnel extraction of enchytraeids, using waterlogged sieves (Graefe 1987, Didden et al. 1995). Depth distribution of enchytraeids was determined by separate extraction of 0-4, 4-8 and 8-12 cm layers of the soil cores. Individuals were grouped into 3 size classes (< 4, 4-8, 8-12 mm). Enchytraeid biomass was calculated using the length-biomass-relation according to Dunger and Fiedler (1989) (1 mm = 0.88 mg fresh weight). The determination of enchytraeids was carried out by U. Graefe, IfAB Hamburg for the sampling in June 2003.

The maximum water holding capacity (% w/ws) of the peat soil was determined by soaking undisturbed soil samples (Ø = 6 cm, depth = 5 cm) from the field for 24 hours, letting them drain on a bed of sand overnight, followed by weighing, drying at 105 °C for 24 hours and re-weighing. During floods, the oxygen content of the water was measured in the field with a WTW oximeter (Oxi340i/Set with a DurOx 325-3 probe). Spearman correlation coefficients were calculated in SPSS for Windows version 11.5.1 for all variables, including the percentage of enchytraeids found in the upper soil layer as well as the number of days between the end of the last inundation and the sampling date. The influence of the factors "site" and "sampling date" on annelid populations was tested by ANOVA (general linear model, univariate). A Chi-square test of the depth distribution of enchytraeids was carried out in Excel following Zöfel (1988).

5.3 Results

5.3.1 Earthworms

Mean earthworm density over the whole sampling period was highest in the gley site (186 ± 240 ind. m^{-2}) and lowest in the peat soil (38 ± 27 ind. m^{-2}; Fig. 5.1). The marsh soil had an intermediate position with a mean earthworm abundance of 87 ± 60 ind. m^{-2}, but mean biomass was somewhat higher than in the gley soil (58 ± 49 vs. 55 ± 63 g m^{-2}; maximum value recorded: 164 ± 236 g m^{-2} in the marsh soil in June 2002). Compared to this, the mean biomass in the peat soil was also very low (7 ± 8 g m^{-2}).

In both Bremen sites, earthworm populations were sharply reduced by flood events. As there was no summer inundation in the marsh site, earthworm populations were stable in 2002, but abundances were low in spring 2003 after the winter inundation (Fig. 5.1a). During the first summer flood day in the peat soil (July 20/02), escaping earthworms from neighbouring sites (the study site being unreachable) were observed to cover distances of several meters, dying on the roads. Afterwards, earthworms were virtually absent from this site, and no recovery took place in the short period before the early winter inundation. At the first spring sampling in May 2003 (the peat being still waterlogged in March), only juveniles in low numbers were found (Fig. 5.1b). In contrast to this, the maximum abundance of earthworms in the gley soil found in August 2002 (Fig. 5.1c, including a maximum biomass of 159 ± 14 g m^{-2} and a very high number of cocoons, 160 ± 69 m^{-2}; data not shown) was hardly reduced by the summer flood. Also biomass stayed at a high level (148 ± 162 g m^{-2}, data not shown), but no cocoons were found. Instead of the expected autumn maximum, a further decline of earthworm abundance before the winter flood occurred, but again a high number of cocoons was found (80 ± 30 m^{-2}, data not shown). After the winter inundation, the abundance was rather low, but still higher than in the other two sites. The age structure of the earthworm communities differed remarkably between the two years of observation: In 2003, in all three sites the proportion of juveniles was much higher than in 2002 (Fig. 5.1).

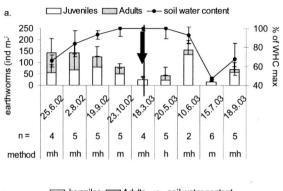

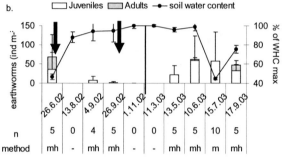

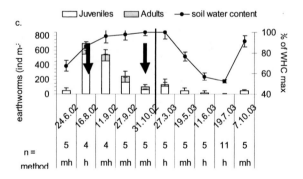

Figure 5.1: Population dynamics of earthworms. Means ± SD of abundances and soil water content (% of max. water holding capacity, n = 5) in (a) marsh soil („Ochtum"), (b) peat soil („Wümme"), (c) gley soil („Gorleben"). Methods: m = mustard extraction, h = hand sorting; mh = combination of both. Arrows indicate flooding events (exact dates only when observed by the authors). Marsh soil: Dec. 01 to Feb. 02, (stagnant rainwater 07-24 to 08-02-2002), 12-15-02 to 02-15-03. Peat soil: Sept. 01, Nov. 01 to Apr 02, 07-20 to 08-21-2002, 10-22-02 to 03-15-03. The site was flooded on some sampling dates and no earthworms could be sampled. Gley soil: Nov. 01 to Feb. 02, 08-18 to 09-11-2002, Nov 02 to Feb. 03

Also during summer drought, the earthworm population dynamics in the two Bremen sites differed considerably from the Gorleben gley site. In the marsh soil, earthworm abundances reached maximum values in June 2003 and only decreased at the climax of the drought period in July 2003, while at the same time in the peat soil, the highest earthworm abundance (with the largest variance) was found (Fig. 5.1a, b). Until September, earthworm abundance stayed at a relatively high level and biomass increased to the maximum value for this site (46 ind. m^{-2} with a distinctly lower variance, 25 g m^{-2}, data not shown), whereas in the gley soil, earthworm abundance had already decreased in May 2003 and no earthworms were present in July.

ANOVA revealed a significant influence of sampling date on abundance and biomass of earthworms (Table 5.2). The number of days since the end of the last inundation correlated positively with abundance and biomass of earthworms in the peat soil (n = 38, r = 0.6 to 0.74, p < 0.001) and negatively with the abundance of earthworms in the gley soil (r = -0.38, p < 0.001, n = 48).

While in the gley soil a higher earthworm abundance was related to a high soil water content over the whole sampling period, (r = 0.43, p < 0.01, n = 39), these parameters were negatively correlated in the peat soil (r = -0.48, p < 0.005, n = 33). In the marsh soil, the correlation tended to be weakly negative (n.s.).

Juveniles and adults were equally abundant in the marsh soil, except for the samplings after the winter flood (March 03) and during summer drought (June and July 03), while in the two other soils, juveniles dominated the populations most of the time (Fig. 5.1). The biomass of juveniles was in the same range or higher than that of adult earthworms in all three soils and on nearly all sampling dates (data not shown).

The efficiency of the mustard extraction was significantly higher in the peat soil (83%) than in the marsh soil (50%) and the gley soil (26%). *L. rubellus* was extracted with a high efficiency from all three soils (62-74 %), just as *O. cyaneum* (75%). For *A. chlorotica* and *O. tyrtaeum*, efficiencies were very low (21% and 31%).

Table 5.2: Significant influence of the factors site and sampling date on population parameters of oligochaetes . * P < 0.05, ** P < 0.01, *** P < 0.001. ANOVA; F-values with significance levels. In brackets: degrees of freedom. *L. rubellus* = abundance of *Lumbricus rubellus*, transformed data (ln [abundance *L. rub*] + 1)

animal group	parameter	site	sampling date	site x sampling date
earthworms	abundance	n.s.	29.8** (23)	n.s.
	biomass	n.s.	4.7** (23)	n.s.
	L. rubellus	n.s.	4.2*** (24)	13.9*** (1)
enchytraeids	abundance	4.4* (1)	11.6** (24)	n.s.
	biomass	19.3** (1)	18.6** (24)	3.2* (2)

Species composition and dynamics. The epigeic earthworm *Lumbricus rubellus* (Hoffmeister) was one of two dominant species in all three sites, the second always being an endogeic earthworm, namely *Allolobophora chlorotica* (Savigny) in gley soil, *Octolasion tyrtaeum* (Savigny) in marsh soil and *Octolasion cyaneum* (Savigny) in peat soil. Additional species recorded as single individuals once or twice were *Lumbricus terrestris* (L.) in gley soil, *Aporrectodea rosea* (Savigny) in marsh soil and *Eiseniella tetraedra* (Savigny) in peat soil.

Abundances of *L. rubellus* were reduced by inundations at all three sites (Fig. 5.2). The short rain water inundation of the shallow ditches in the marsh site in July 2002 did not affect this species, but after the winter inundation, it was absent, just as in the peat soil after the summer inundation. In the gley soil, abundances were moderately reduced after the summer flood and only later reduced to zero. During summer drought, this species throve well in the marsh and the peat soil, but was reduced to zero again in the gley soil (Fig. 5.2). ANOVA revealed a significant influence of date and date x site, but only a trend for site alone (Table 5.2).

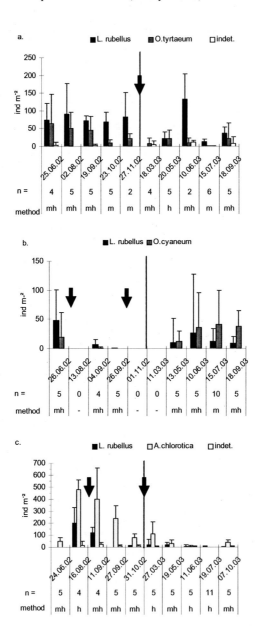

Figure 5.2: Earthworm species dynamics in three flooded grassland sites. Mean values (n as indicated in Fig.) ± SD of abundances in a. marsh soil („Ochtum"), b. peat soil („Wümme"), c. gley soil („Gorleben "). Arrows indicate flooding events; for exact dates see Fig. 5.1. In the peat soil, the site was flooded on some sampling dates and no earthworms could be sampled.

The dynamics of the two *Octolasion* species in the two Bremen sites were comparable in 2002 and in spring 2003: The populations were stable, but reduced by flooding. During summer drought, *O. tyrtaeum* in marsh soil underwent a distinct decline, whereas *O. cyaneum* in peat soil became increasingly dominant with increasing drought (Fig. 5.2a, b). *A. chlorotica* in gley soil was found to be in diapause, i.e. curled up in soil between the deepest grass roots, not reacting to the mustard suspension, being only extractable by hand-sorting on most sampling dates. It had hardly changed its density after the summer flood; it disappeared during summer drought in 2003, but recovered quicker than *L. rubellus* after the first rainfall in autumn (Fig. 5.2c).

5.3.2 Enchytraeids and other oligochaetes

Over the whole sampling period, mean abundance and biomass of enchytraeids were highest in the marsh soil (12,000 ind. m^{-2}; 0.048 g m^{-2}) and lowest in the gley soil (4,900 ind. m^{-2}; 0.017 g m^{-2}). The peat soil took an intermediate position with 8,100 ind. m^{-2} and 0.032 g m^{-2}.

In the peat soil, population dynamics (abundance and biomass) of earthworms and enchytraeids were in parallel ($r = 0.52$, $P < 0.001$, $n = 36$). In the other soils, there was no correlation between the total abundances of earthworms and enchytraeids. Similar to the earthworms, the enchytraeids had stable populations during the wet year 2002 in the unflooded marsh soil. The summer flood events in the other sites led to a collapse in the enchytraeid populations (Fig. 5.3b and 5.3c). Before winter inundation, abundance recovered to 30% (peat soil) and 100% (gley soil) of the early summer abundance. During the first days of the early winter flood in the peat soil and the inundation of the ditches in the marsh soil (due to an artificially raised groundwater level and rainfall) in November 2002 only single individuals were found alive (Fig. 5.3a, b). In the dry summer 2003, abundance and biomass rose to maximum values in the marsh and the peat soil, followed by a decline in September, whereas the population in the gley soil had already decreased in June and reached minimum values in July and October (Fig. 5.3). ANOVA revealed a significant influence of sampling date, site and sampling date x site on abundance and biomass of enchytraeids (Table 5.2).

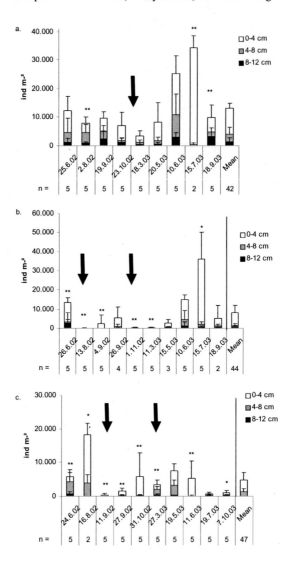

Figure 5.3: Population dynamics and vertical distribution of enchytraeids. Mean values and standard deviations (n as indicated in Fig.) of abundances in different soil depth. a. marsh soil („Ochtum"), b. peat soil („Wümme"), c. gley soil („Gorleben "). Arrows indicate flooding events; for exact dates see Fig. 5.1. Mean of all sampling dates excluding samplings in stagnant water after flooding events (i.e. for marsh soil: 27/11/02; for peat soil: 13/8/02, 1/11/02 and 11/03/03, for gley soil: 11/09/02). Asterisks indicate significant deviations from the mean depth distribution (Chi square test, * => P < 0.05; ** => P < 0.01, degrees of freedom= 2).

A negative correlation between soil water content and enchytraeid abundance was found for the gley soil in general ($r = - 0.45$, $p < 0.01$, $n = 40$) as well as for the smallest enchytraeids in all three soils (< 4 mm; gley soil: $r = - 0.38$, $p < 0.05$, $n = 35$, peat soil: $r = - 0.48$, $p < 0.001$, $n = 33$; marsh soil: $r = - 0.39$, $p < 0.05$, $n = 30$).

The number of days since the end of the last inundation was positively correlated with abundance and biomass of enchytraeids in the peat soil ($n = 46$, $r = 0.6$ to 0.74, $P < 0.001$; additional figure available on request). In both peat and gley soil, it correlated negatively with the percentage of enchytraeids in the upper four centimetres (peat: $r = - 0.51$, $P < 0.001$, $n = 30$; gley: $r = -0.32$, $P < 0.05$, $n = 38$).

11/09/02). Asterisks indicate significant deviations from the mean depth distribution (Chi square test, $* => P < 0.05$; $** => P < 0.01$, degrees of freedom= 2).

Depth distribution. Most enchytraeids were found in the upper 4 cm of the soil, the highest mean portion in the peat soil (Fig. 5.3). Means were calculated without inclusion of sampling dates directly after or during flooding events when only single enchytraeids were found. The Chi-square test revealed numerous deviations from this mean distribution for sampling dates when the soil was waterlogged, during drought or when mass reproduction in the upper layer took place.

Species composition. The enchytraeid communities of the three study sites differed considerably. While in the peat soil, *Cognettia glandulosa* (Michaelsen) was eudominant with 89 % of the individuals, it accounted for only 10 % in the gley soil, next to *Henlea ventriculosa* (Udekem) (54 %) and *Fridericia bulboides* (Nielsen & Christensen) (16%). The marsh soil was dominated by undeterminable juveniles of the genus *Fridericia* (49 %, probably also *F. bulboides*, as 3 % of the mature individuals belonged to this species). The only species occurring at all three sites were *C. glandulosa, F. bulboides,* and *H. perpusilla* (Friend).

The marsh and the peat soil were also inhabited by the aquatic oligochaete *Lumbriculus variegatus*, always occurring in the upper 4 cm. After flooding, its dominance was considerably higher in both soils, and in the peat soil it was the only oligochaete at all (data not shown). Aquatic tube worms (Tubificidae; *Rhyacodrilus* spec.) were recorded as single individuals in all three soils.

5.4 Discussion

5.4.1 General population dynamics and species composition

In the course of the whole sampling period, the marsh soil was settled most densely by annelids. With its microrelief offering unflooded refuge sites and depressions where water contents always remained at a relatively high level, it seemed to match best their environmental needs. The generally low abundance and biomass of earthworms in the peat soil can be explained by the low pH, while enchytraeids are known to be abundant in such acidic, SOM-rich soils (Didden 1993). In the gley soil, earthworms probably profited from the higher soil pH, at least at moderate humidity. The lower SOM content and WHC but also competition by earthworms may have limited enchytraeid populations. When comparing the three sites, the overall relationship between earthworm and enchytraeid abundances is negative, as described in the literature (Schaefer and Schauermann 1990, Górny 1984). However, because of the impact of repeated hydrological stress, the dynamics of the two groups were in parallel most of the time. Presumably, this stress is also reflected by the very high juvenile/adult ratio in earthworms at most sampling dates in all three sites. The mustard extraction worked well for earthworms staying near the surface (*L. rubellus* as epigeic species and *O. cyaneum* because of lack of oxygen in the waterlogged peat soil), while it failed for endogeic species, especially during the diapause of *A. chlorotica*.

According to the indicator system for annelid communities (Graefe and Schmelz 1999, Beylich and Graefe 2002), all dominant enchytraeid species found in this study are K strategists, explaining why the disturbance of summer floods had a lasting effect on their densities.

5.4.2 Effects of inundation

As earthworms have various physiological adaptations to overcome the lack of oxygen under water (Bouché 1977, Edwards and Lofty 1977, Lee 1985), several earthworm species including *L. rubellus* may survive up to several weeks in flooded soil (Pizl 1999, Ausden et al. 2001, Zorn et al. 2004a). Enchytraeids extracted from soil thrive well in oxygen-rich water for several days, but only when temperatures are kept between 6 and 10 °C and when there is not too much decaying organic matter in the sample (Graefe, pers. comm.). This decaying organic matter leads to a depletion of oxygen during an inundation in the field, especially at high temperatures. In the last week of the Wümme summer flood, at a water temperature of 21°C, oxygen saturation was about 8% in the shallow water covering on the SOM rich peat soil. Comparably low oxygen saturations were measured repeatedly during flooding simulation experiments in laboratory microcosms at a temperature of 15°C (Chapter 7). The absence of annelids after the summer flood in 2002 presumably was due to this low oxygen content of the water. The following early winter inundation did not allow the populations to recover. Also in other studies (Faber et al. 2000, Ausden et al. 2001, Zorn et al. 2004b) annelid populations were sharply reduced by flooding events.

The relatively weak impact of the summer flood on earthworms in the gley site can be explained by the moderate temperatures, implying a higher oxygen content of the flood water, as well as by the shorter duration of the Elbe flood. Besides, *A. chlorotica* was already in summer diapause when the flood came, so its reduced metabolism may have ensured its survival. Also Zorn et al. (2004b) reported that the species was hardly affected by flooding in the field.

The quick drainage of the gley soil allowed for a quick recovery of annelid populations after flooding. The negative correlation between water content and enchytraeid abundance in all three soils reflects their general sensitivity towards flooding. Most enchytraeids did not survive flooding, not even at moderate winter temperatures, as the sampling in the first days of the winter inundation on the peat and the marsh soil in autumn 2002 showed. However, their populations recovered quicker than those of earthworms.

The deviations from the normal depth distribution of the enchytraeids reflect the dynamics of hydrological conditions: In the marsh soil, the constant humidity allowed a relatively stable vertical distribution of enchytraeids, whereas in the gley soil, the depth distribution deviated from the mean distribution on nearly every sampling date, due to changing hydrological extremes. In the upper soil layers, an exchange with air is best, thus, this layer has the highest oxygen content during an inundation (Healy 1987). This is probably why enchytraeids concentrated there in all samplings during the wet period, most of all in the peat soil.

One survival strategy of earthworms when their habitat is flooded is to seek refuge in dryer sites, like the ridges in the marsh site and the summer dyke in the gley site. When the soil drains again, they recolonise the lower parts. However, fleeing earthworms easily become prey to birds (Ausden et al. 2001), and therefore predation also may have impacted on abundance of *L. rubellus*. On the peat site, a large area (more than 600 ha, WWF 2002) is flooded and distances are too large to be covered by smaller earthworms like *L. rubellus*, as their rate of migration is too slow (Hoogerkamp et al. 1983, Curry and Boyle 1987, Marinissen and van den Bosch 1992, Ausden et al. 2001, Klok et al. 2004). Thus, the new generation there can only have hatched from cocoons. Obviously these are resistant to inundation, for more than 6 winter months and also for shorter periods at high temperatures. This is evident from the high number of earthworm cocoons preceding the summer flood in the gley soil and the successful recovery afterwards.

L. rubellus is an r strategist (Wallwork 1983) which adapted to the life in frequently flooded sites with a lower mature weight (Klok et al. 2004). The early maturation probably helped *L. rubellus* to recover quicker than *O. cyaneum* from the summer flood in the peat site, namely in the same year in autumn. However, obviously most of the *L. rubellus* juveniles did not survive the early winter flood. The building up of the new population from the few remaining cocoons in 2003 was slower than that of *O. cyaneum*, and the latter dominated in summer.

5.4.3 Effects of drought

In the peat soil, earthworm and annelid populations clearly profited from the long inundation-free period in 2003. Even in the driest phase, high water capacity of the peat permitted high annelid abundances. No collapse of the earthworm population could be observed, and enchytraeids even had a mass-hatching in the peat as well as in the marsh soil. In this spatially well-structured site, the contrasting dynamics of earthworms and enchytraeids at the climax of drought can only be explained by a greater mobility of the earthworms. As earthworms are known to move with a changing gradient of humidity (Ausden et al. 2001), it is very likely that they did so in June 2003 when the ridges in the marsh soil dried up. Thus, the high abundances of *L. rubellus* found in the ditches then could have been aggregated populations from the drier ridges. However, the drought climax in July 2003 also reduced the species there. *O. tyrtaeum*, a typical wetland species, seems to be sensitive towards drought because its abundances were distinctly lower in 2003 as compared to the previous year.

The gley soil with its comparatively low water holding capacity drained much more rapidly than the other soils, and drought already started in spring. Furthermore, the drought stress was probably most severe in this subcontinental climate and in this soil. The consequences were decreasing annelid abundances with increasing time after inundation and a concentration of enchytraeids in deeper layers, like described in literature (Springett et al. 1970, Briones et al. 1997). Furthermore, the impact of the summer drought in 2003 on the earthworms was more pronounced than that of the summer flood in 2002.

5.4.4 Conclusions

The results of this field study show that earthworms and enchytraeid populations are severely reduced by inundations. Sampling data of annelids in dynamic habitats like floodplains have to be interpreted carefully. Earthworm abundances and population structure differ strongly in time as influenced by contrasting hydrological conditions. To obtain mean population values, sampling has to be done across several years.

Where controlled flooding is applied at a large scale, recolonisation by annelid worms from unflooded refuge sites is not likely to occur. Therefore, the time interval between two flooding events should exceed the development time from cocoon to adult of the earthworm and enchytraeid species present. To maintain viable populations of annelids, especially in spring when they serve as food for wetland birds, a new inundation in the recovery period (2-3 months for enchytraeids and about half a year for earthworms) should be omitted or kept short. In periods of excessive drought, water management should keep soil water contents at a tolerable level by artificially raising the groundwater table. In any case, the flooding regime should take into account the respective site conditions, i.e. water holding capacity and SOM content of the soil.

5.5 Acknowledgements

We are grateful to members of our department for assistance with the field and laboratory work as well as to Ulfert Graefe, IfAB Institut für Angewandte Bodenbiologie (Hamburg) for determining the enchytraeid worms and giving helpful comments on the manuscript. Klemens Ekschmitt provided additional information to his earthworm population study in the "Niedervieland". We also thank the Senator für Bau, Umwelt und Verkehr Bremen, the NLfB (Niedersächsisches Landesamt für Bodenforschung, especially Bernd Kleefisch and Hubert Groh), the WWF (Worldwide Fund for Nature, especially Gunnar Oertel) and the BUND (Bund für Umwelt und Naturschutz Deutschland, especially Birgit Olbrich) for cooperation in selecting study sites and getting permissions to take samples.

6. Efficiency of mustard for extracting earthworms from floodplain meadows

Abstract: In preparing field investigations, recent literature about different earthworm extraction methods (hand-sorting, formalin, mustard, electrical octet method) was reviewed. In the course of two years, earthworms were extracted six times with a suspension of green mustard seed from three floodplain meadows differing in soil type: a marsh soil, a peat soil and a gley soil. The efficiency of the method was checked by hand sorting on the same plot.

The mustard suspension extracted between 0 and 100% of the earthworms present on a sampling plot, depending on site and earthworm species. Extraction efficiency was significantly higher in the peat soil (83%) than in the marsh soil (50%) and the gley soil (26%). The efficiency correlated negatively with total earthworm abundance. The epigeic *Lumbricus rubellus* was extracted with a high efficiency from all three soils (62-74 %), and also 75 % of the endogeic *Octolasion cyaneum* were extracted from the peat soil, was extracted. For the two other endogeic species, *Allolobophora chlorotica* in the gley soil and *Octolasion tyrtaeum* in the marsh soil, efficiencies were very low (21% and 31%, respectively). A significant positive correlation between soil water content and extraction efficiency was only found for *O. cyaneum*.

The results show that the mustard extraction method for earthworms, combined with an efficiency control through hand sorting, is suited to extract earthworms from a representative area. However, the extraction efficiency strongly depends on site (soil type) and earthworm species (life forms). Therefore, extraction data should always be shown separately for different earthworm species, especially in site comparisons. Also soil humidity, water table depth and the state of the earthworms (diapause or active state) influence the extraction efficiency. As these problems also occur in using the formalin method, the further development of the harmless and thus preferable mustard method is recommended. In using the pure active substance contained in all mustard types, allyl isothiocyanate, a standardisation should be possible.

6.1 Introduction

The most simple and most reliable method for the extraction of earthworms, including inactive specimens as well as cocoons, is hand-sorting. However, anecic (deep-burrowing) species are missed because they escape to deeper layers when digging is started (Ehrmann and Babel 1991, Chan and Munro 2001), and physical disturbance of the soil is required by this time-intensive and laborious method. In spite of the recent development of a time-limited soil sorting procedure (Schmidt 2001), the method can only be applied for small volumes. To enhance the probability to detect all species present on a site and to obtain representative data on abundance, biomass and population structure of earthworms, hand sorting is combined with a chemical extraction applied over a greater area.

The standard chemical extraction with formalin, introduced by Raw (1959; see also ISO 2004) does not only imply health hazards for the applying person, but also has severe side effects on vegetation and soil fauna (Gronstol et al. 2000). The application of 0.4 % formalin solution decreased vegetation cover, shoot and root biomass as well as soil respiration and the total abundance of Collembola and Nematoda. These effects lasted for the whole vegetation period (Eichinger 2004). Nagel (1996) found a reduction of microbial activity after formalin application lasting for 6 to 8 weeks, followed by a compensatory effect (higher microbial activity for more than 10 weeks). Also household detergent, although sometimes efficient in expelling earthworms (Gronstol et al. 2000), may severely damage the animals' health (East and Knight 1998).

Gunn (1992) was the first to use English mustard paste as an alternative expellant for earthworms on park grassland, finding the method to be more than twice as efficient as formalin and other chemical extractants. In the meantime, the mustard method has already entered official working procedures as a suitable alternative to formalin (Eidgenössische Forschungsanstalten 2002, ISO 2004). Notably, not the processed mustard (as in Gunn's paper), but ground seeds should be used to prepare a suspension of acceptable efficiency. This is probably the reason why different authors found slightly lower efficiencies for the mustard method compared to the formalin method (Gronstol et al. 2000, Tiefenbrunner and Tiefenbrunner 1993), as well as compared to hand-sorting (East and Knight 1998). But also Fründ and Jordan (2004) found lower efficiencies, using both prepared mustard and ground seeds. In the study of Emmerling (1993), the mustard method had the highest efficiency compared with formalin, vinegar and the electrical octet method. Chan and Munro (2001) showed that mustard is 67% more efficient than formalin in extracting the anecic *Anisochaeta* sp. However, many soil ecologists prefer formalin, doubting the efficiency of the mustard extraction method. This study aims to contribute to the discussion with some data on the efficiency of mustard suspension for the extraction of different earthworm species and age stages from three soil types under changing hydrological conditions.

I assumed that the extraction efficiency of mustard suspension (i) would be in the same range or higher than that of the formalin method described in the literature; (ii) would be higher for epigeic than for endogeic earthworms; (iii) would differ between adults and juveniles, (iv) would differ for abundance and biomass of earthworms and (v) would differ between soil types.

6.2 Material and methods

Earthworms were sampled from three floodplain meadows six times in the course of two years. For the main vegetation and soil characteristics of the study sites, see Table 3.1. As two of the study sites are situated in nature reserves, the formalin method was ruled out. Another alternative, the electrical octet method, had been studied by Blanken-Mittendorf (1990) in comparable floodplain meadows, with numerous technical problems when the soil water content was too high (which occurs frequently after inundations or heavy rains) or too low. Thus, the mustard extraction method was chosen and its efficiency was assessed by hand sorting on the same plot.

100 g of ground green mustard seed (Develey, Germany, Art.-No. 2098) were allowed to swell for at least 2 hours in 1 l of water. This swelling time is indispensable for the efficiency of the method, as the total failure of a first trial with a suspension that had only swelled for 15 minutes had shown. Directly before application, this suspension was diluted to 3 x 10 l in plastic canisters, giving a 0.33 % concentration.

To prepare the extraction sites (normally 5 replicates per date and site, see Fig. 6.1), vegetation was cut short and the removed plant material was searched for earthworms. The suspension was distributed evenly inside a wooden frame (50 x 50 cm) until the soil was waterlogged and a second time after 10 min. (maximum: 10 l of suspension). For 30 minutes (beginning after the first application), appearing earthworms were collected in a plastic bag with about 1 cm of water. For control, one quarter of the same plot (25 x 25 x 20 cm, a depth always reaching the groundwater table in these sites) was dug with a spade and hand-sorted. Earthworm biomass was estimated by a volumetric method (displacement of water in a scaled vessel, 1 ml being equivalent to 1.064 g fresh biomass, Dunger and Fiedler 1989). The abundances were converted to ind. m^{-2} for both methods and combined to give the total density of earthworms.

Relative to this number (= 100%), the efficiency of the mustard extraction method was calculated as the proportion of earthworms extracted with mustard suspension. Using SPSS version 12.0., the efficiency for adults and juveniles, abundance and biomass of the different species was compared with a T-test. Spearman correlation coefficients were calculated for extraction efficiency, earthworm abundance and soil water content. As the preconditions for an ANOVA were not given, the site influence on extraction efficiency was tested with a Kruskal-Wallis H-test followed by a Tamhane PostHoc test.

6.3 Results

The mustard suspension extracted between 0 and 100% of the earthworms present on a sampling plot, depending on soil type and earthworm species. The highest mean efficiency (83%) was obtained on the peat soil site, compared to 50% in the marsh soil and 26% in the gley soil (Fig. 6.1), giving an overall mean of 53%. The soil type had a significant influence on extraction efficiency (Kruskal Wallis H = 22.995; p < 0.001). The pair-wise tests showed significant differences between marsh and peat soil (p < 0.01) as well as between peat and gley soil (p < 0.001), while for the marsh and gley soil comparison, only a trend was found (p = 0.060).

Mustard sampling in the gley soil in June 2002 during a relatively dry period (soil water content: 68% of WHC_{max}) totally failed, while the highest efficiencies (100% in the peat soil, 72% in gley soil) were found at high soil water contents (90-100% of WHC_{max}). However, a high soil water content did not always imply a high extraction efficiency (marsh soil in March 2003: 37%; peat soil in May 2003: 57%; gley soil on both sampling occasions in September 2002: 10%; Fig. 6.1). The efficiency correlated negatively with total earthworm abundance (for all three sites together: [R = -0.42, p < 0.001, n = 67]; for the marsh soil site: [R = -0.514, p < 0.05, n = 21] In the other two sites, no such correlation was found. There was no difference in extraction efficiency between earthworm abundance and biomass, nor between juveniles and adults (abundance and biomass) in all three soils (T-tests not significant; Fig. 6.2).

The collected species reacted differently to the mustard suspension. The epigeic *Lumbricus rubellus* was extracted with a high efficiency from all three soils (62-74 %, Fig. 6.2), and also the endogeic *Octolasion cyaneum,* occurring in the peat soil, was extracted with an efficiency of 75%. For the two other endogeic species, *Allolobophora chlorotica* and *O. tyrtaeum*, efficiencies were very low (21% and 31%, respectively). However, the only significant difference in extraction efficiency for two earthworm species occurring in one site was found between *L. rubellus* and *A. chlorotica* in the gley soil (T = 4.98, p < 0.05). *A. chlorotica* were curled up in a diapause stage at most sampling dates and seldom showed a reaction to the direct touch during hand sorting. They only began to move when they were wetted by the water in the storage bag.

A significant correlation between soil water content and extraction efficiency was only found for *O. cyaneum* (R = 0.89, p < 0.05, n = 5).

In addition to earthworms, numerous other soil- and surface-dwelling animals reacted immediately to the mustard suspension: enchytraeids, collembolans, beetles (Staphylinidae and others) and their larvae, some mobile dipteran larvae, spiders, mites, cicadina, and, on one sampling occasion in the peat soil, a leech.

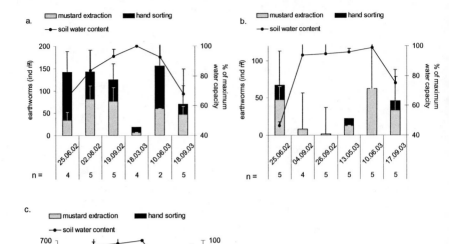

Figure 6.1: Efficiency of the mustard extraction method in different soils with changing soil water contents. a. marsh soil, b. peat soil, c. gley soil. Mean individual numbers of earthworms with SE as revealed by the mustard extraction (grey) and additional hand sorting (black). Soil water content in % of the maximum water holding capacity.

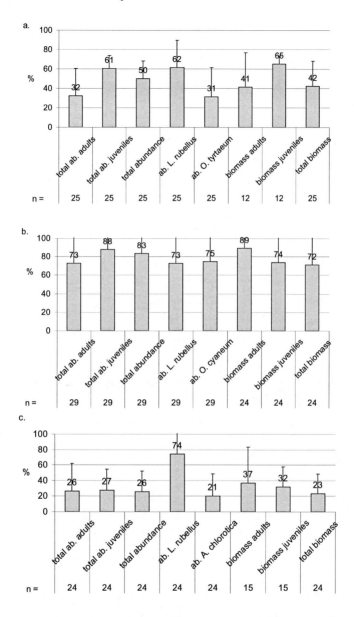

Figure 6.2: Efficiency of the mustard extraction method for abundance, biomass, different age stages and species of earthworms. Means and SE of replicates from all sampling dates. a. marsh soil (n = 25 except for biomass adults and juveniles: n = 15); b. peat soil (n = 29 except for all biomass values: n = 24); c. gley soil (n = 24, except for biomass adults and juveniles: n = 15). ab. = abundance

6.4 Discussion

The mean extraction efficiency of the mustard extraction method determined in the present study is in accordance with Lawrence and Bowers (2002): They detected an average of 61 % for earthworm abundance, 62% for their biomass and 84.5% of the species richness on 40 sampling sites differing in land use type. They did not find any differences in efficiency across different soil and habitat types.

Filser, Mommertz, Ackermann, and Ehrengruber (unpublished) reported a higher extraction efficiency with higher proportions of clay and silt, as earthworms are wetted more efficiently by the mustard suspension when there are less coarse pores. Probably for the same reason, compaction of the soil also had a positive influence on extraction efficiency. Unfortunately, data on the distribution of grain sizes of the peat soil in the present study are not available. The marsh soil indeed has a higher clay content and a higher extraction efficiency than the gley soil. However, the difference in extraction efficiency between the soils probably rather depends on the different endogeic earthworm species found there (see below).

The positive correlation between soil water content and extraction efficiency for *O. cyaneum* can be explained by the high water table in the peat. Oxygen deficiency forces the earthworms to concentrate in the upper layer of the soil where oxygen contents are higher and where there is likely to be direct contact with the mustard. Besides, the 100% efficiency with the highest soil water contents could be due to a bias of the method, as hand sorting is difficult in waterlogged soil and, in spite of diligent and skilful work, smaller earthworms especially are easily overlooked.

As juvenile earthworms have a thinner skin and react faster, it seems very likely that the mustard extraction works better for them than for adults, as revealed in the marsh and the peat soil. However, as they are smaller, the distance to the surface is more difficult to cover than it is for adults. This could be the reason why Lawrence and Bowers (2002) found a slightly higher effectiveness of the mustard extraction for adult earthworms, while in the present study, no difference between the age stages was found.

The total failure of the method on the gley soil site in June 2002 was probably due to the relatively dry soil conditions causing the diapause of *A. chlorotica*. In general, this endogeic species is rather inactive, while the epigeic *L. rubellus* is more active, resulting in a relatively good extraction efficiency of the mustard for this species in all three sites. These results are corroborated by the findings of Fründ and Jordan (2004), where no *A. chlorotica* and *Aporrectodea caliginosa* (another endogeic species) were extracted with mustard, while the efficiency was 100% for *L. rubellus*, *L. castaneus* and *L. terrestris*. Also Chan and Munro (2001) reported very low extraction efficiencies for the endogeic *Aporrectodea trapezoides*: 35.5 % with the formalin method and even lower efficiencies with different mustard suspensions. The subsequent hand-sorting revealed that most of the *A. trapezoides* were dead. However, the authors also mention the more limited contact of this species with the mustard because the worms mainly live in horizontal burrows in the soil without direct contact with the surface. The same explanation is given by Lawrence and Bowers (2002) concerning the low efficiencies for the endogeic *O. tyrtaeum*. For the anecic *L. terrestris*, the same authors detected an average efficiency of 70%, which was above the mean and can be attributed to the good flow of the mustard along the vertical burrows of this species. No statement about anecic (deep-burrowing) earthworms can be made from the present study as they do not occur on the study sites. A possible problem with the method is that, when epigeic and anecic species are more easily extracted with mustard suspension and endogeic species are missed, the apparent dominances of the earthworm species extracted could be misleading. However, with the formalin method the same problem occurs; endogeic species are only extracted efficiently by hand sorting or a combination of the methods (Ehrmann and Babel 1991).

The type of seeds and the mode of preparation are of importance: green mustard seeds are preferable to yellow seeds and cold preparation of the suspension over hot preparation (Filser, Mommertz, Ackermann, and Ehrengruber, unpublished). Moreover, the right concentration has to be chosen. With a 0.1 % mustard suspension, earthworms left the soil very slowly, and the efficiency was lower than that of the formalin method, while a 0.33% mustard flour suspension had the same extraction efficiency as the formalin method (Högger 1993). In a laboratory experiment, a concentration of 1% did not extract more earthworms than the 0.33% concentration. It only made them leave the soil faster, rendering their collection more difficult when they occur in high abundances in the field. Besides, an increased secretion of mucus was observed and interpreted as a stress symptom (Filser et al., unpublished). Also the death of juveniles due to high mustard concentrations can reduce sampling efficiency (Chan and Munro 2001).

In contrast to formalin, mustard (if ready or as ground seeds) can only be prepared as a suspension and not as a solution. As a consequence, its efficiency is strongly dependent on physical soil characteristics such as pore size (Tiefenbrunner and Tiefenbrunner 1993). Another problem is the varying efficiency of different mustard types, rendering the comparison of different studies difficult (Gronstol et al. 2000). Both problems can be solved by the application of the pure active substance contained in all mustard types, allyl isothiocyanate (AITC). Zaborski (2003) suggested that it might be a favorable alternative to formalin expulsion for sampling earthworms. He determined an optimal concentration of 100 mg l^{-1} AITC in expelling *Lumbricus terrestris* and *Aporrectodea tuberculata*. This substance is a major degradation product of glucosinolates from tissue of plants belonging to the genus *Brassica* that are also used to control soil-borne plant pests. With reduced soil moisture and higher temperatures, it is transformed more rapidly, i.e. within a range of 22 h to 35 h (Borek et al. 1995). Thus, this is no explanation for the low efficiencies of the mustard extraction with dry soil conditions in the present study, as the whole extraction only takes 0.5 h. Besides, soil organic carbon correlated negatively with the half-live of allyl isothiocyanate (Borek et al. 1995). Obviously, this had no influence on the present results, as the highest efficiencies were recorded in the peat soil which had the highest content of C.

To conclude, the mustard extraction method for earthworms is suited to extract earthworms from a larger area than would be possible with hand-sorting only. However, hand sorting is necessary to determine the efficiency of the method which depends strongly on earthworm species (life forms). Therefore, in site comparisons extraction data should always be shown separately for different earthworm species. As the formalin method also requires calibration by hand-sorting and, thus far, there is no evidence for lower efficiency of mustard as compared to formalin, the harmless mustard method should be preferred by ecologists, also outside nature reserves. However, in view of the different extraction efficiencies of different mustard types, standardisation by using allyl isothiocyanate is necessary. The possible side effects of this substance should be investigated prior to standardisation.

6.5 Acknowledgements

I warmly thank Juliane Filser, Susanne Mommertz, Généviève Ackermann and Annette Ehrengruber for providing unpublished results from their experiments with mustard suspension. I am also grateful to Alexander Bruckner, Tamara Coja and Eva Eichinger, BOKU Vienna, for the exchange of literature. My special thanks go to the companies Kühne (Hamburg) and Develey (Unterhaching) for the free offers of mustard flour.

7. Earthworms and potworms (Lumbricidae, Enchytraeidae) mobilise soil nutrients from flooded grassland

Abstract: The role of earthworms and potworms (Oligochaeta: Lumbricidae, Enchytraeidae) in terrestrial-aquatic nutrient exchanges of flooded grassland ecosystems was investigated in a laboratory microcosm experiment with flooded soil cores differing in soil type (peat, marsh, gley soil). During 72 hours, there was a high, site-dependent mobilisation of nutrients from all soil cores to the nutrient-poor river water. The earthworms generally increased the mobilisation of nutrients or had no effect. Enchytraeids alone never had a significant effect, but some lumbricid effects were only significant in their presence. On the peat soil, concentrations of ammonium and phosphorus in the flood water were increased in the presence of the earthworm *Octolasion cyaneum*. The presence of *Lumbricus rubellus* led to a mobilisation of nitrate and phosphate from marsh soil and of nitrate and nitrite from gley soil. In the peat and the marsh soil, the correlation of ammonium and water pH was increased in the presence of enchytraeids. As annelid casts are known to be rich in inorganic N and P, they are a possible source of the mobilised nutrients. Further indirect effects of annelids mediated by burrowing, selective grazing and the fostering of nitrification are discussed. It is recommended to exclude annelid-rich floodplain sites from controlled inundation to prevent undesired nutrient release into the rivers.

Key words: Lumbricidae; Enchytraeidae; wetland; inundation; terrestrial-aquatic element exchange; nutrient mobilisation

7.1 Introduction

Ecosystem research in terrestrial and aquatic systems has developed as two separate subdisciplines. To be able to deal with biogeochemical phenomena across large landscapes where these ecosystems are connected, these subdisciplines have to be merged (Grimm et al. 2003). The present study concentrated on the interface of a terrestrial and an aquatic system, one of the so-called biogeochemical "hot spots" (McClain et al. 2003). Controlled inundation of river floodplains in Western Europe (rivers Rhine, Elbe and others) is practised at a large scale. While the retention of excessive water prevents inundations in settled areas, the flooding of nutrient-poor grassland can reduce the concentration of undesired nutrients and pollutants in the flood water (Haslam et al. 1998). In addition to this, wetland ecosystems are promoted as a habitat for special plant communities and birds. Soil invertebrates are an important food resource for wading birds, and high densities, especially of earthworms, are a desired goal of wetland management (Brandsma 1997; Ausden et al. 2001).

The aim of our laboratory experiments was to investigate the role of terrestrial invertebrates in nutrient exchange between soil and river water in three flooded grassland ecosystems in Northern Germany. Field investigations in 2002 and 2003 (Chapter 5) revealed that the dominant group amongst soil meso- and macrofauna in the study sites was the annelids: earthworms (Lumbricidae) and potworms (Enchytraeidae). The annelid populations were reduced by extensive flooding, especially during the warmer season, like observed in other studies (Pizl 1999, Faber et al. 2000, Chapter 4). However, their role in the biogeochemical "hot moment", the first few days of a flooding event (McClain et al. 2003), may be considerable. Annelid worms foster the growth of micro-organisms (Wolters 1988, Daniel and Anderson 1992, Hedlund and Augustsson 1995) which use nutrients for building up their biomass. Thus, an indirect nutrient immobilising effect can be assumed and has been observed in some cases (e.g. Nagel et al. 1995). The guts and casts of earthworms have been identified as denitrification "hot spots" (Svensson et al. 1986, Scheu 1987, Parkin and Berry 1994), and could therefore form an important pathway for the loss of N from soil and water. However, numerous studies in terrestrial ecosystems have shown that inorganic N and P are released from decomposed plant litter, digested microorganisms, and excrements in the presence of annelids (earthworms: Sharpley and Syers 1976, Scheu 1993, Haynes et al. 2003; enchytraeids: Williams and Griffiths 1989, Briones et al. 1998). Based on these observations, the following hypotheses were raised:

1. During a flooding event in grassland, earthworms and enchytraeids enhance the mobilisation of ammonium, nitrate and phosphorus from soil into the flood water.

2. Nutrient release and the role of annelid worms in these processes differ with respect to soil type.

7.2 Material and methods

We aimed to simulate flooding of undisturbed soil in microcosms in the presence and absence of annelid worms. The experiment was set up in such a way that the effect of earthworms and enchytraeids alone as well as the interaction effect of both groups together could be compared to an animal-free control. Each treatment was replicated 5 times.

7.2.1 Study sites

Soil samples and annelids were taken from three regularly winter-flooded meadows along rivers in Northern Germany with differing soil types (for soil parameters, vegetation and hydrological history of these sites see Table 7.1). The sampling was carried out in May 2002. All three sites are mowing meadows without any application of fertilisers or herbicides. The gley soil site is subjected to various contaminants from the river flood-water (dioxins, heavy metals, locally also PAH and furanes; NLfB 1997).

The water (about 10 l) was taken with a big flask from the rivers which regularly inundate the meadows. For water sampling, a place upstream the inundation area was chosen. Water was taken from freely flowing part of the river to minimize the content of suspended sediment. The water was filtered the same day and cooled at 15°C until the start of the experiment, then it was filtered again and water parameters were determined.

Table 7.1: Vegetation and soil characteristics of the study sites

Main data sources: [a] Erber 1998; [b] B. Olbrich, pers. comm.; [c] B. Schuster, unpublished data; [d] Janhoff 1992; [e] G. Oertel, pers. comm., [f] Kleefisch and Knes 1997; [g] own measurements / observations; [h] www.wetteronline.de/reuro.htm, 21.10.2004.

Study site	Nature reserve "Ochtumpolder bei Brokhuchting", Bremen	Nature reserve "Borgfelder Wümmewiesen, Bremen	Soil monitoring site of the State Lower Saxony, Gorleben/Elbe
Soil type	marsh soil[a]	peat (10-15 cm) on river sand[d]	river gley soil[f]
pH (CaCl$_2$)	4.8 [a]	4.6 [c, d]	5.5[f]
N$_t$ (%)	0.7 [a]	2.3 [c, d]	0.7[f]
C$_{org}$ (%)	6.2 [a]	29.1 [d]	9.5 [f]
C/N	9.1 [a]	14.8 [c, d]	13.4 [f]
SOM (%)	> 10 [a]	27.8 [c, d]	9.3 [f]
Field capacity (vol.-%)	64 [a]	67 [d]	64 [f]
Water holding cap. (mass %)	64 [a]	76 [g]	52 [f]
Special characteristics	oxide concretions [a]	peat mixed with river sediments [d]	contamination (see text) [f]
River (tributary to stream)	Ochtum (Weser)	Wümme (Weser)	Elbe
Climate	oceanic	oceanic	subcontinental
Water management, inundation interval	backwater flooding every winter from December to February [b]	backwater flooding in winter, episodic summer floods at high water levels [d]	no management, natural inundations every second winter, episodic summer floods [f]
Inundations before and during the study period	Dec. 01 to Feb. 02, (rainwater 07-24 to 08-02-2002), 12-15-02 to 02-15-03 [g]	Sept. 01, Nov. 01 to Apr 02, 07-20 to 08-21-2002, 10-22-02 to 03-15-03 [g]	Nov. 01 to Feb. 02, 08-18 to 09-11-2002, Nov 02 to Feb. 03 [g]
Plant association	Phalaridetum arundinaceae [a, g]	Caricetum gracilis (Calthion) [d, g]	Phalaridetum arundinaceae, Glycerietum fluitantis [f, g]

Table 7.1 (continued): Vegetation and soil characteristics of the study sites

Study site	Nature reserve "Ochtumpolder bei Brokhuchting", Bremen	Nature reserve "Borgfelder Wümmewiesen, Bremen	Soil monitoring site of the State Lower Saxony, Gorleben/Elbe
Root horizon	10 cm [a]	13 cm [d]	7 cm [f]
Agricultural use	mowing 1-2 times/year; sometimes pasture in autumn (not in years of study) [b]	mowing 1-2 times/year [e]	mowing 2-3 times/year [f]
precipita-tion (mm y^{-1}) 2002	1070 [h]	1070 [h]	852 [h]
2003	607 [h]	607 [h]	427 [h]
sun hours 2002	1523 [h]	1523 [h]	1501 [h]
sun hours 2003	1885 [h]	1885 [h]	2005 [h]

7.2.2 Annelid worms used for the experiment

As the experiment claims to simulate natural conditions, the three types of soil cores were inoculated with site-characteristic annelid communities and not with the same set of species.

The epigeic *Lumbricus rubellus* (Hoffmeister) was chosen as a representative earthworm as it is one of the dominant species at all three study sites. However, after repeated inundations, *L. rubellus* was no longer found on the peat site, so the endogeic *Octolasion cyaneum* (Savigny) was collected there and used for the inoculation of the peat soil microcosms. Enchytraeids were extracted as black-box-communities from each soil, as their identification under the microscope easily destroys their fragile bodies. Previous field investigations showed that the dominant enchytraeid species were *Henlea ventriculosa* (Udekem) in the gley soil, *Cognettia glandulosa* (Michaelsen) in the peat soil and *Fridericia* (*sp.*) juv. - probably *F. bulboides* (Nielsen and Christensen) - in the peat and the marsh soil.

Abundance of annelids was chosen to simulate values of previous field surveys on all three sites. For earthworms, the simulated density of 600 ind. m^{-2} is within a maximum range for the gley soil, whereas for the other soils, such a density was never observed during the field investigations (maximum abundance: gley soil 696 ± 66 ind. m^{-2}, marsh soil 156 ± 91 ind. m^{-2}, peat soil 67 ± 52 ind. m^{-2}). The high simulated density was chosen to be sure to detect a possible effect of the animals. The simulated enchytraeid density of 6000 ind m^{-2} was within the lower range of mean field abundances (gley soil 4900 ± 5500 ind. m^{-2}, marsh soil 13,000 ± 11,000 ind. m^{-2}, peat soil 8100 ± 11,000 ind. m^{-2}). However, at favourable soil moisture conditions, a much higher maximum abundance of enchytraeids (20,000 to 50,000 ind. m^{-2}) occurred in all three soils (Chapter 5).

7.2.3 Course of the experiment

Soil samples were collected on the 19th and 20th of May 2003 with a root drill (diameter = 8 cm) and transported to the laboratory in plastic bags. The cores were defaunated by freezing at –18°C and thawing (3 times). The vegetation was cut as short as possible and remaining litter was removed carefully to prevent nutrient release from decaying plant material. The upper 6 cm with a mean fresh weight of 254 ± 24 g (SD) were put as undisturbed soil cores into PE flasks of the same diameter (8 cm, height = 13 cm). These microcosms were installed in a room at a permanent temperature of 15°C.

Before the inoculation of microcosms, earthworms were set on wet filter paper overnight, and their fresh weight was determined thereafter. Enchytraeids were extracted from fresh soil samples using the cold wet funnel technique (Graefe 1987, Didden et al. 1995) and grouped in three size classes (< 4 mm, 4.1 - 8 mm, 8.1 - 12 mm). Even proportions (by number) of enchytraeid size classes were given to each replicate and treatment. Enchytraeid biomass was calculated using the length-biomass-relation according to Dunger and Fiedler (1989) (1 mm = 0.88 mg fresh weight).

Earthworms were given one week to adapt to the new environment. Enchytraeids were put onto the soil cores only five days later to prevent a change in numbers by death and/or reproduction (especially the rapid way of reproduction by fragmentation that was observed on *C. glandulosa*). Thus, the enchytraeids only had two days of adaptation. Then the microcosms were carefully flooded with 400 ml of filtered water from each river, giving a water-to-soil ratio (vol:fw) of 1.6. This amount of water was chosen to simulate the mean height of flood water observed in the field (about 5 cm). The short duration of flooding (only 72 hours as against several weeks to months in the field) was chosen to guarantee the survival of the annelid worms, as the role of living individuals in nutrient release and not of their dead tissue was to be investigated.

The oxygen content of the water was measured daily during the experiment with a WTW oximeter (Oxi340i/Set with a DurOx 325-3 probe), the electrical conductivity with a WTW Cond340i probe and the pH with a Knick 766 Calimetric probe. Nutrient concentrations of river water (kept in a separate open flask in the same room) and microcosm flood water after the 72 hour incubation period were analysed by photometric analyses: the sodium-salicylate-method (Hach test) for ammonium, the sulphanilamide method (Dr. Lange test) for nitrite, the cadmium reduction method (Hach test) for nitrate, and the ammonium-molybdate-method after digestion with peroxodisulfate (Dr. Lange test) for total phosphorus.

At the end of the experiment, the soil cores were taken out of the water and searched for earthworms by hand-sorting. The surviving earthworms were weighed again with the same procedure as used at the start of the experiment. All soil cores were put in wet funnels to extract surviving enchytraeids, but also to control if enchytraeids or nematodes occurred in the other treatments. The identification of enchytraeids was provided by U. Graefe, IfAB Hamburg. A portion of soil (10 g dry weight) was used for the measurement of pH (CaCl$_2$).

As the water in this kind of microcosms also reaches the sides and the bottom of the soil cores, the contact surface of soil and water is enhanced and could cause an unnatural nutrient mobilisation. To determine the extent of this possible influence, microcosms closing tightly around the soil cores were constructed with steel rings so that the water was only in contact with the soil surface. Water nutrient concentrations were then compared with those in the PE flasks containing the same amount of soil.

For comparison of nutrient mobilisation factors, some results from a field experiment on the marsh soil site are presented. There, nutrient concentrations in flooded mesocosms directly inserted into the soil were measured after 1.5 hours of flooding (same water-to-soil ratio as in the laboratory experiment).

7.2.4 Statistics

All variables were tested for normal distribution with the Kolmogoroff-Smirnov test and for homogeneity of variances with the Levene-Test (SPSS, version 11.5). If the preconditions were satisfied, an ANOVA followed by a Bonferroni Post Hoc-Test were used to compare the treatments, otherwise the Kruskal-Wallis H-Test, followed by the Tamhane Post Hoc-test was used. Spearman correlation coefficients were calculated for all variables, but only significant correlations are mentioned or shown in figures.

7.3. Results

7.3.1 Survival of annelid worms

While all endogeic earthworms (*O. cyaneum*) survived and gained weight (0.3 ± 0.19 mg SD per set of 3 worms), half of the *L. rubellus* died during the simulated inundation. Only about 3 % of the enchytraeids were recorded at the end of the experiment (Table 7.2), comprising individuals of all dominant genera recorded in the field (*Cognettia, Henlea, Fridericia, Enchytraeus*). To exclude an influence of dying annelid worms on the measured parameters, the difference between weight of living *L. rubellus* individuals at the beginning and at the end of the experiment (thus: the weight lost by dying individuals, 0.1 to 1.4 g per microcosm) and the number of enchytraeids that were not recovered (26 to 30 per microcosm) were included as variables in the calculation of Spearman correlation coefficients.

Table 7.2: Surviving animals after 72 h of flooding

Site	earthworms (out of 3 per microcosm)	enchytraeids (out of 30 per microcosm)
peat soil	3 ± 0.0 (*O. cyaneum*)	0.6 ± 1.3
marsh soil	1.3 ± 1.3 (*L. rubellus*)	0.4 ± 0.7
gley soil	1.6 ± 1.0 (*L. rubellus*)	0.3 ± 0.7

7.3.2 General development of flood water chemistry

During the 72 hours of flooding, the oxygen saturation in the flood water as well as the water pH decreased considerably in all soils and treatments when compared to the initial river water measurements (Table 7.3). Nutrients were mobilised in all treatments including the animal-free control (Fig. 7.1). Comparing the three soils, the highest mobilisation of ammonium occurred in the peat soil (333-fold increase compared to the initial river water). Total P was mobilised by a factor of 112 and 357 from the marsh and the gley soil, respectively. The latter also released high amounts of ammonium (140-fold increase) and nitrate (95-fold increase) as well as nitrite (concentrations of 1 to 7 mg l^{-1} compared to a nitrite concentration of < 0.001 mg l^{-1} in the river water). Measurements of nitrite in water samples taken randomly from microcosms with peat and marsh soil cores only revealed negligible concentrations (< 0.04; <0.001 mg l^{-1} NO_2^- in river water).

No significant difference between nutrient concentrations or other parameters in PE flask and steel ring microcosms were found (Table 7.4).

Table 7.5 shows the factors of increase of nutrient concentration in flood water on marsh soil after 1.5 hours of inundation in field mesocosms compared to those after 72 hours in laboratory microcosms. In all treatments of both experiments, all measured nutrients were released from soil, and the factor was highest for ammonium. For nitrate, the mobilisation was even higher after 1.5 hours than after 72 hours.

Table 7.3: Oxygen saturation and pH in terrestrial-aquatic microcosms with different soil types at the end of the laboratory experiment.

Soil	treatment	O_2 satur. (%)	pH water	pH soil
peat soil	river water	90	7.7	-
	L	21 ± 9	5.9 ± 0.2	5.1 ± 0.1
	E	29 ± 7	6.1 ± 0.1	5.1 ± 0.1
	L+E	14 ± 6	6.1 ± 0.2	5.0 ± 0.1
	C	20 ± 5	6.1 ± 0.1	5.2 ± 0.1
marsh soil	river water	80	7.4	-
	L	25 ± 15	6.4 ± 0.2	4.9 ± 0.1
	E	23 ± 6	6.2 ± 0.1	4.8 ± 0.1
	L+E	26 ± 3	6.5 ± 0.2	4.9 ± 0.1
	C	29 ± 9	6.3 ± 0.2	4.8 ± 0.1
gley soil	river water	98	8.0	-
	L	48 ± 7	6.6 ± 0.1	5.6 ± 0.1
	E	61 ± 4	6.5 ± 0.1	5.4 ± 0.1
	L+E	46 ± 9	6.5 ± 0.2	5.5 ± 0.1
	C	67 ± 3	6.6 ± 0.2	5.5 ± 0.1

Measurements after 72 h. For river water values: n = 1. For microcosms flood water: n = 5 (mean values of each treatment with standard deviations). L = lumbricids, E = enchytraeids, C = control.

Table 7.4: Comparison of flood water parameters in two types of microcosms with peat soil cores after 24 hours

Parameter/Microcosm	PE flask	Steel ring
pH water	5.5 ± 0.04	5.4 ± 0.05
oxygen saturation (%)	39 ± 7	34 ± 12
electrical conductivity ($\mu S\ cm^{-1}$)	662 ± 51	620 ± 54
water absorbed by soil core (ml)	133.4 ± 11	184.4 ± 4.5
NH_4^+-N (mg l^{-1})	6.3 ± 3.8	7.55 ± 2.06
NO_3^--N (mg l^{-1})	12.8 ± 4.9	12.2 ± 7.5
total P (mg l^{-1})	0.19 ± 0.1	0.25 ± 0.16
pH soil	4.81 ± 0.04	4.85 ± 0.09

Mean values (n = 5) and standard deviations of animal-free microcosms. The nitrate values are much higher than in the described experiment because the soil cores were taken at the beginning of April (just after the end of a winter inundation). At this time of the year, mineralisation starts due to rising temperatures, while plants are not ready for the take-up of nitrate.

Table 7.5: Mobilisation of nutrients from marsh soil to river water

Mobilisation after...	NO_3^--N		NH_4^+-N		total P	
	earthw.	contr.	earthw.	contr.	earthw.	contr.
...1.5 h in the field	2.9	2.4	25	21	4.3	4.4
...72 h in the laboratory	1.6	1.8	100	47	100	80

Values indicate the factor of increasement of nutrient concentration when river water and mesocosm water are compared. "earthw." = treatment with increased earthworm abundance (field: 110 ind. m-², laboratory: 600 ind. m-²; both L. rubellus), field: n = 3; laboratory: n = 5; "contr." = control treatment with reduced earthworm populations (field, n = 4) or without earthworms (laboratory, n = 5).

7.3.3 Effects of annelid worms

In the flood water of the peat soil treatment, significantly higher concentrations of ammonium and total phosphorus were recorded in the presence of *O. cyaneum* compared to the animal-free control. When earthworms and enchytraeids were present, the maximum value for total phosphorus was even higher (Fig. 7.1a, Table 7.6). The soil pH was slightly lower in the presence of annelid worms (Table 7.3, Table 7.6). In the marsh soil, only the treatment with earthworms and enchytraeids showed significantly higher concentrations of phosphorus and nitrate when compared to the animal-free control (Fig. 7.1b, Table 7.6).

In microcosms with gley soil cores, the high concentration of nitrite in the flood water was increased in the presence of earthworms and enchytraeids (Fig. 7.1c, Table 7.6). For all other nutrients (NH_4-N, NO_3-N and total P), no significant differences were observed between the control and the treatments with annelids.

In the field experiment (Table 7.5), some replicates failed and the number of the remaining was too low to show any significant effects of earthworms.

7.3.4 Correlations

In the peat soil treatments with enchytraeids, ammonium concentrations and pH of the flood water were highly positively correlated (E: $r = 0.98$, L+E: $r = 0.96$; $P < 0.01$ for both treatments; Fig. 2a). The same phenomenon was observed in the marsh soil (E: $r = 0.82$, not significant; L+E: $r = 0.98$, $P < 0.01$; Fig. 2b).

The weight of dead *L. rubellus* individuals displayed a highly negative correlation with nitrate and total phosphorus concentrations for the gley soil ($r = -0.90$, $P < 0.05$), while the number of dead enchytraeids was negatively correlated with nitrate (marsh and gley soil), ammonium, phosphorus and the pH of water and soil (only marsh soil) ($r = -0.63$ to -0.76, $P < 0.05$). For the peat soil, there was no significant correlation between the number of dead enchytraeids and any of the measured parameters.

Table 7.6: Results of ANOVA and Kruskal-Wallis testing of the effects of annelids on parameters of flood water and soil of different microcosm treatments.

Parameter	Soil	F /H value	Posthoc
NH_4^+ water	peat soil	F = 4.04*	L > C*
NO_3^- water	marsh soil	F = 3.9*	L+E > C*
total P water	peat soil	F = 6.4**	L > C*
total P water	marsh soil	H = 12.3**	E < L+E*
			L+E > C*
O_2 saturation water	peat soil	F = 3.8*	E > L+E*
O_2 saturation water	gley soil	F = 14.3***	L < E*
			L < C*
			L+E < C*
soil pH	peat soil	F = 6.9**	L < C*
soil pH	marsh soil	F = 4.95*	L > E*

Results of ANOVA (F values) and Kruskal-Wallis-Tests (H values) with Post-Hoc tests for differences between treatments Bonferroni and Tamhane, respectively). Asterisks show significant differences of treatments with worms compared to the control (* at the 0.05 level, ** at the 0.01 level, *** at the 0.001 level). n.s = not significant. L = with Lumbricids, E = with Enchytraeids, L + E = with both groups, C = animal-free control

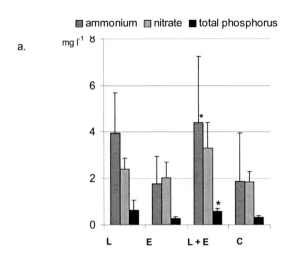

a.

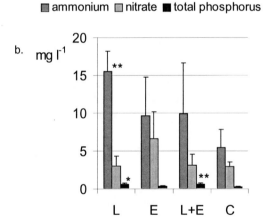

b.

Figure 7.1: Mean concentrations of nutrients in the flood water of terrestrial-aquatic microcosms in treatments with and without annelid worms. n = 5; **a. marsh soil, b. peat soil.** L = with lumbricids, E = with enchytraeids, L+E with both groups, C = animal-free control. Asterisks show significant differences of treatments with worms compared to the control (* at the 0.05 level, ** at the 0.005 level). Concentrations in river water at the start of the experiment (before the contact with the soil cores): 0.03 mg l^{-1} NH$_4^+$, 1.0 mg l^{-1} NO$_3^-$, 0.005 mg l^{-1} total P.

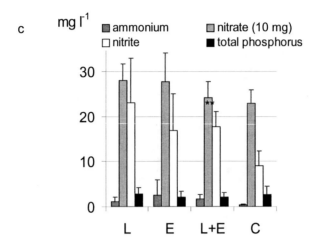

Figure 7.1 (continued): Mean concentrations of nutrients in the flood water of terrestrial-aquatic microcosms in treatments with and without annelid worms. n = 5; **c. gley soil.** L = with lumbricids, E = with enchytraeids, L+E with both groups, C = animal-free control. Asterisks show significant differences of treatments with worms compared to the control (* at the 0.05 level, ** at the 0.005 level). Concentrations in river water at the start of the experiment (before the contact with the soil cores): 0.03 mg l^{-1} NH_4^+, 1.0 mg l^{-1} NO_3^-, 0.005 mg l^{-1} total P.

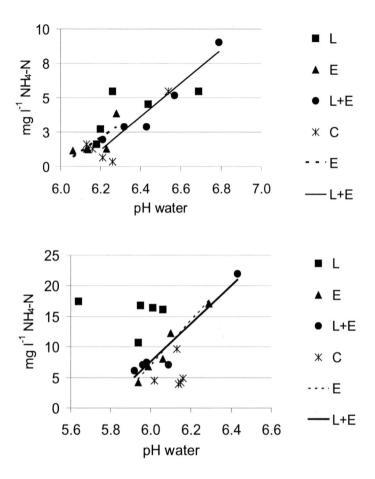

Figure 7.2: Correlation of ammonium and pH in the flood water of microcosms in treatments with and without annelid worms. a. peat soil, b. marsh soil; L = with lumbricids, E = with enchytraeids, L+E with both groups, C = animal-free control

7.4 Discussion

7.4.1 Methodological remarks

The generally high nutrient mobilisation, also from animal-free controls, was not expected. Wetlands are known to remove especially nitrate from inundation water (Haslam et al. 1998). It is taken up by plants or, probably to a greater extent, transformed into N_2 and N_2O by denitrification under anoxic conditions in the water-saturated or inundated soil (Well et al. 2002, Hefting et al. 2003). As the edge surface area apparently had no effect on nutrient mobilisation from the controls, it can be assumed that the deep-freezing and thawing of soil cores caused the high amounts of nutrient release in the controls. Although freezing is an effective method for defaunation (Kampichler et al. 1999), nutrient dynamics are influenced by the breakdown of microbial, plant and animal tissue (Mack 1963, Williams and Griffiths 1989). Kampichler et al. (1999) only found a slightly higher NH_4^+-N-leaching, but in some cases a very high release of NO_3^--N from soil monoliths defaunated by deep-freezing and thawing compared to unfrozen controls. In a field experiment on the marsh soil site, the mobilisation of ammonium and phosphorus were lower than in the laboratory experiment presented here (factors 20-30 and factor 4, respectively).

A lower oxygen saturation of the flood water in the presence of animals was expected, but not found. The relatively high variability of soil mass per microcosm may have obscured such effects.

A possible effect of enchytraeids on nitrification at high densities as they occur in the field can be assumed regarding the effects observed at the simulated lower density in the current study. Although the experiment was kept short, most enchytraeids obviously did not survive flooding. The death of the animals after some days of flooding, even at lower air and water temperatures, has already been observed in the field (Chapter 5). Earthworms vacate flooded soil (Ausden et al. 2001, K. Ekschmitt, pers. comm.; original observations). However, in wetlands with a great extent like the "Borgfelder Wümmewiesen" (peat soil), no escape is possible, and enhanced earthworm activity while fleeing could reinforce their effect on nutrient release.

7.4.2 Effects of annelid worms

The increased ammonium and total phosphorus concentrations in the presence of earthworms (and enchytraeids) can be explained by nutrients released from excrements of the animals. Earthworm casts are normally produced on the soil surface (80 to 460 mg g^{-1} live worm per day for *L. rubellus;* Shipitalo et al. 1988) and hence would be directly exposed to flood water. Fresh earthworm casts are known to be rich in NH_4^+-N, NO_3^--N, and total phosphorus (Sharpley and Syers 1977, Parkin and Berry 1994). In enchytraeid faeces (Wolters 1988) and probably also in earthworm casts, nutrients seem to be occluded by the compact structure.

However, in water-saturated soil, like in the present experiment, this protecting effect is reduced and inorganic N is leached (McInerney and Bolger 2000). In all these studies, the release and transport of nutrients from above to below (by percolating or runoff water) induced by earthworms is described. A fresh aspect of the present study is the opposite direction of nutrient displacement in stagnant water, from below to above. Ammonium contents of fresh casts tend to be particularly high, but decrease over time due to nitrification (Syers et al. 1979, Scheu 1987, Martin and Marinissen 1993, Haynes et al. 2003). As phosphatase activity in earthworm casts is high, organic bound P is transformed into inorganic P and can therefore be easily leached (Satchell and Martin 1984, Scheu 1987, Krishnamoorthy 1990). Furthermore, enchytraeids use earthworm casts as a food resource, promoting the decomposition of enclosed litter (Zachariae 1967). This could lead to the enhanced nutrient mobilisation when both groups of annelids are present.

An increase of N_{min} leaching up to a factor of 2.0 was recorded in the presence of the earthworm *O. lacteum* (Scheu 1993, 1995) and explained by the burrowing activity of the endogeic earthworm. However, the burrowing could supply new surfaces for ammonium to adsorb and thus reduce the concentrations in flood water. Therefore, it is more likely that ammonium will be released from casts and secretes under flooded conditions. Like other mesofauna organisms, enchytraeids are known to play a regulating role in mineralisation processes through their selective grazing on micro-organisms (Dash and Cragg 1972). They can increase the leaching of dissolved organic nitrogen, ammonium, and phosphorus (Williams and Griffiths 1989, Briones et al. 1998). No such effect was evident in the present experiment. However, the positive correlation of ammonium and water pH in the presence of enchytraeids and *L. rubellus* in the peat and the marsh soil could possibly be explained by the sorption of H^+ ions on micro-organisms or the tissue provided by the (dead) animals. Further support to this is given by the fact that the degree of this correlation decreased as the amount of SOM decreased in the three studied soils (peat > marsh > gley; see Table 7.1).

In many experiments and field studies in terrestrial ecosystems, an indirect effect of annelid worms, e.g. on nitrification by fostering the involved micro-organisms, has been observed (earthworms: Weiß 1994, Helling 1997, enchytraeids: Williams and Griffiths 1989, Filser and Krogh 2002). In the current study, this trend could only be observed for the marsh soil where NO_3^- concentrations in the flood water were increased in the presence of earthworms and enchytraeids together. Additionally, the negative correlations of NO_3^- with the number of dead enchytraeids hints at a role of the living animals in nitrification. In peat, the low pH possibly inhibited nitrification in general (see Table 7.1). The accumulation of nitrite in the gley soil could have been caused by the enchytraeids fostering the micro-organisms involved in the first step of nitrification.

7.4.3 Recommendations for water management

Wetlands that are known to be densely populated by annelid worms should be spared from controlled flooding, at least during the vegetation-free period, to prevent undesired nutrient release into the river water during short-term inundations of some days. As during long-term inundations of several weeks to months, earthworms are killed, this practice is counterproductive to its aim to deliver food for wading birds (Ausden et al. 2001). In wastewater treatment reed beds, earthworms have been observed to clean the gravel substrate by translocating clogging material from the interstices to the surface, thus prolonging the life time of the reed bed (Davison et al. 2005). As the purpose of reed beds is the treatment of wastewater, often to remove nutrients, their inoculation with earthworms should be considered carefully, taking into account the counterproductive nutrient-mobilising effects of the animals.

7.4.4 Research needs

Given that ammonium and nitrate are not very persistent in water, repeated measurements during the first 24 hours of an inundation would reveal the temporal dynamics of nutrient release in the presence and absence of annelid worms. This would also ensure that the effect of living annelids could be investigated, especially that of enchytraeids. A flooding experiment with varying annelid densities would be worthwhile to reveal possible density-dependent effects, especially concerning pH-dependent nitrification. In order to determine whether an enchytraeid effect is produced by active (moving, feeding, and casting) individuals or by their dead tissue, observations of enchytraeids in waterlogged soil should be carried out. A comparison of the effect of different life forms of earthworms (endogeic and epigeic species) can only be made by observing them in the same soil type. With this in mind, an experiment with *L. rubellus* on peat soil would be the easiest way to reveal possible differences compared to *O. cyaneum*. Also the role of dead earthworm tissue in nutrient release should be investigated, seen the long inundation times in the field, causing the dead of the animals.

To verify the validity of the measured effects in this laboratory experiment, a field experiment is necessary. The ideal mesocosms should allow a lateral water movement like it is brought about by rising flood water. Nutrient concentrations should not only be measured in the water, but also in soil, ideally in those higher, shallow and calmer zones where solved and suspended material is normally deposited during floods.

7.5. Acknowledgements

This study was carried out at the Department of General and Theoretical Ecology in the UFT Centre of Environmental Research and Technology at the University of Bremen. We are grateful to several members of our department, the Department of Aquatic Ecology and the Institute of Environmental Process Engineering (IUV) for their practical assistance in the field and laboratory work. Ulfert Graefe, Institute for Applied Soil Biology (IfAB, Hamburg) identified the enchytraeids. We also thank the Senator für Bau, Umwelt und Verkehr Bremen, the NLfB (Niedersächsisches Landesamt für Bodenforschung), the WWF (Worldwide Fund for Nature) and the BUND (Bund für Umwelt und Naturschutz Deutschland) for co-operation in selecting study sites and getting permissions to take samples. Our special thanks go to Tom Headley as well as to two anonymous referees for helpful comments on the manuscript.

8. Temporal variation in nutrient release from a flooded peat soil depending on earthworms and enchytraeid density

Abstract: Earthworms and enchytraeids are known to mobilise nutrients from soils, which is an undesired process when floodplains are inundated and released elements are added to the river water. Based on results from an earlier experiment, we hypothesised that (i) the correlation of ammonium and pH is related to enchytraeid density; that (ii) the release of nutrients in the flood water is increased by enchytraeids; that (iii) in the presence of the earthworm *Octolasion cyaneum*, the enchytraeid effect is more pronounced, and that (vi) enchytraeid effects would be most pronounced at intermediate density.

In microcosms, flooding was simulated in soil cores taken from a peat soil, inoculated with enchytraeid communities (*Fridericia* spp. and *Cognettia glandulosa*) in 5 different densities (0; 2,000; 4,000; 8,000 and 16,000 individuals m^{-2}) in the presence and absence of one adult *O. cyaneum*. After 24 hours, oxygen saturation, electrical conductivity and pH of the flood water were measured and nutrient concentrations determined photometrically. All individuals of *O. cyaneum* survived the experiment, while only 11% of the enchytraeids were found back. Probably as a consequence of this high mortality or inactivity of the animals when flooded, no measurable effects of enchytraeids on parameters or correlations between them were found, except for density-dependent effects in single replicates with freshly extracted enchytraeid populations. In the presence of *O. cyaneum*, various parameters differed and correlated with each other. Further differences were recorded for soil cores taken from the field at different times of the year. The impact of soil animals on this process varies over time, which requires further research when it should be taken into account in controlled water management.

8.1 Introduction

Soil fauna, in particular oligochaetes, are known to have a large, mostly indirect, impact on nutrient turnover. In wetland soils, this impact has a special significance because mobilised nutrients are taken away by flood water. The mobilising influence of oligochaetes on nutrients in terrestrial systems is well known (earthworms:

Haynes et al. 2003, Ruz-Jerez et al. 1992, Sharpley and Syers 1977; enchytraeids: Briones et al. 1998, Schrader et al. 1997, Sulkava et al. 1996, Wolters 1988). In the presence of the earthworm *Octolasion lacteum*, mineral N in soil leachate was increased (Scheu 1993). In an experiment with enchytraeids, concentrations of mineral N and phosphorus in leachate were increased and their biomass was positively correlated with NH_4^+ concentration in soil leachate. This was explained by enchytraeid grazing on microorganisms in which these nutrients formerly were immobilised (Williams and Griffiths 1989). From studies on mycorrhizal mycelium, a density-dependent effect of microarthropod grazing (here: Collembola) has been observed (Finlay 1985). Also the effect of sediment-dwelling macrozoobenthos on nutrient mobilisation from lake sediments has been studied, with contradictive results: While Fukuhara and Sakamoto (1987) recorded enhanced concentrations of ammonium in the lake water and an accelerated release of phosphorus when tubificids and/or chironomid larvae were present, Lewandowski and Hupfer (2005) measured lower concentrations of total P compared to animal-free controls. However, the role of terrestrial invertebrates in terrestrial-aquatic matter exchange has been neglected in previous studies. In a first laboratory experiment with flooded soil cores from different soil types, the mobilisation of ammonium and total phosphorus from a peat soil was significantly increased in the presence of the earthworm *Octolasion cyaneum* (Savigny). This effect was even more pronounced in the simultaneous presence of enchytraeids. In treatments with enchytraeids (in presence as well as in absence of earthworms), ammonium release into the flood water and pH of the water were highly positively correlated. The same phenomenon was observed in a marsh soil with another enchytraeid community and the earthworm *Lumbricus rubellus* (Hoffmeister) (Chapter 7). This hints at a significant role of the animals in stimulating the activity of non-pH-dependent nitrifyers. Based on these results, we hypothesised that (i) the correlation of ammonium and pH is related to enchytraeid density; that (ii) the release of nutrients in the flood water is increased by enchytraeids; that (iii) in the presence of *O. cyaneum*, the enchytraeid effect is more pronounced, and that (iv) enchytraeid effects would be most pronounced at intermediate density.

8.2 Material and methods

Soil samples and annelids were taken from a mown sedge meadow (*Caricetum gracilis;* Calthion) in the nature reserve Borgfelder Wümmewiesen in Bremen, Northern Germany. The reserve is flooded naturally every winter and occasionally also in the warmer season by the river Wümme, a contributary of the Weser/Lesum. There is a certain man-made regulation of flooding regimes with respect to flooding time. The soil is a peat layer mixed with fluvial deposits (10-15 cm), on river sand. The soil had a pH ($CaCl_2$) of 4.6, a maximum water holding capacity of 67 vol. %, a C/N ratio of 14.8 and a soil organic matter (SOM) content of 27.8 % (B. Schuster, pers. comm.).

Twenty-five waterlogged undisturbed soil samples (d = 8 cm, depth = 5 cm, fresh weight: 210 ± 24.7 g, water content: 80 g g^{-1}) were taken on March 3, 2004 and transported to the laboratory in plastic bags, together with turfs from which the enchytraeids could be extracted later on. The turfs were stored at 15°C in the laboratory and wetted regularly. As one replicate failed, five additional soil cores were taken on April 5 when the soil was not waterlogged anymore (water content: 66 g g^{-1}). The intact cores were defaunated by repeated freezing at -18°C and thawing. After thawing the soil cores, the vegetation was cut as short as possible and most of the litter was removed. The cores were incubated for 5 days at 15°C. Treatments with different enchytraeid densities (2000, 4000, 8000 and 16,000 individuals m^{-2}) with and without the earthworm *Octolasion cyaneum* were set up. Over five weeks, every week one replicate consisting of all treatments was set up, beginning on April 13 (10 treatments, 5 replicates). Each time, enchytraeids were extracted from soil samples taken from the turf stored in the laboratory with the cold wet funnel extraction, using waterlogged sieves (Didden et al. 1995, Graefe 1987). The animals were grouped in three size classes (< 4 mm, 4.1 - 8 mm, and 8.1 - 12 mm). According to the availability of animals, equal portions of these size classes were added to every microcosm. Calculated enchytraeid biomass according to the length-biomass-relation (Dunger and Fiedler 1989: 1 mm ^= 0.88 mg fresh weight) was 0.8 mg fresh weight for the lowest densities of 2,000 ind. m^{-2} and 5.2 mg for the highest densities (16,000 ind. m^{-2}). The dominating genera *Cognettia* and *Fridericia* (occurring in a relation of 1:1) as well as the rare *Marionina argentea* and some single *Enchytraeus* spp. could be distinguished under the binocular and were added in equal portions to the microcosms. The zero-density served as an animal-free control. Adult individuals of the endogeic earthworm *Octolasion cyaneum* had been collected on the study site in autumn 2003 and kept in the laboratory (at 15°C) in boxes with their natural soil and alder leaves (*Alnus glutinosa*) as food during winter. Before the inoculation, they were set on wet filter paper overnight, and their fresh weight (0.74 ±0.29 g) was determined. Earthworms and enchytraeids were given one day to adapt to the new environment before the microcosms were flooded with filtered water from the river Wümme for 24 hours.

The oxygen content of the flood water in the microcosms was measured with a WTW oximeter (Oxi340i/Set with a DurOx 325-3 probe), the pH with a Knick 766 Calimetric probe, and the electrical conductivity with a WTW Cond340i probe. Water samples of 10 ml were taken from the microcosm flood water with a pipette and filtered (for determination of ammonium-N and nitrate-N) or left unfiltered (for determination of total phosphorus). The concentrations of these nutrients were analysed by photometric analyses (Dr. Lange method, CADAS 100 photometer). Soil cores were hand-sorted to find back the earthworms that were kept to be used for the next replicate. The wet funnel method was used again (including the cores of the animal-free control) to investigate the number of surviving enchytraeids that were determined under a microscope. A portion of soil (5 g dry weight) was used for the determination of soil pH ($CaCl_2$).

Statistics. Statistical analysis was carried out with the help of the SPSS Statistical Package for the Social Sciences version 12.0.1. When the preconditions (normal distribution, homogeneity of variances) were met, treatments with and without *O. cyaneum* as well as treatments with soil cores taken in March or April were compared with a T-Test, otherwise with a Mann-Whitney-U-Test. The influence of enchytraeid density on measured parameters in either presence or absence of *O. cyaneum* was tested with a 2-way-ANOVA. Spearman coefficients were calculated for all measured variables; quadratic regressions with the curvefit function were made for the different enchytraeid densities, for single replicates as well as for the whole data set.

8.3 Results

8.3.1 Survival of annelids

Out of the 1,500 enchytraeids used for the inoculation, only 156 (i.e. about 11%) were found alive in the end of the experiment. Next to single individuals of the genus

Enchytraeus and several undeterminable juveniles, the proportion of the dominating genera *Cognettia* and *Fridericia* was still 1:1. All earthworms had survived. Besides, several nematodes were found in the wet funnel extraction for enchytraeids (0-4 ind. per pot; mean: 1.34 ind.). No nematodes or enchytraeids were found in control treatments.

8.3.2 Mobilisation of nutrients

In general, high amounts of phosphorus and ammonium were mobilised from the flooded soil cores; pH and oxygen saturation of the flood water were lowered after 24 hours of flooding. For the soil cores taken in March, there was no difference between river water and flood water from microcosms in electrical conductivity, and the cores reduced nitrate concentrations in the river water by about 50%. The additional soil cores taken in April (n = 10) showed a distinctly different mobilisation pattern. Independent of the worms' influence, pH and ammonium concentration of the flood water were lower than on soil cores taken in March, while the conductivity, the total phosphorus as well as the nitrate concentration were higher. Nitrate values in the flood water on soil cores taken in April were more than ten times higher than those in the flood water on soil cores taken in March and about 500-600% of those in the river water. Variances of all variables (except electrical conductivity) measured in the flood water on April-soil cores were distinctly higher than those of treatments with soil cores taken in March. Thus, this additional replicate was excluded from all calculations.

8.3.3 Influence of enchytraeid density

The enchytraeid density had no influence on any of the measured parameters (ANOVA and Kruskal-Wallis H-Test results, Fig. 8.1). For all treatments without earthworms together (n = 20), there was a negative correlation of oxygen saturation with concentrations of total phosphorus (r = -0.77**, Fig. 8.2a.) and a positive correlation with NO_3^--N (r = 0.48*), while the latter was positively correlated with the soil pH (r = 0.6**; not shown in Figures).

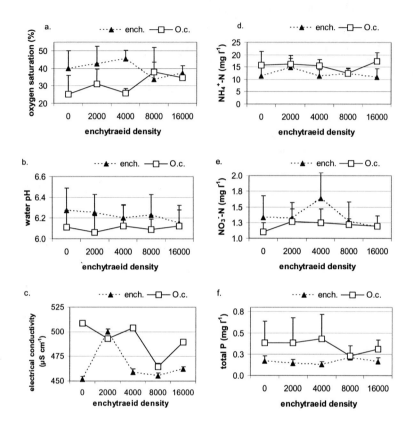

Figure 8.1: Water parameters in treatments with different enchytraeid densities (ind. m⁻², mean values of n = 4 ± SD). Triangles: enchytraeids only; squares: with the earthworm *O. cyaneum*. a. oxygen saturation, b. water pH, c. electrical conductivity, d. ammonium-N, e. nitrate-N and f. total phosphorus.

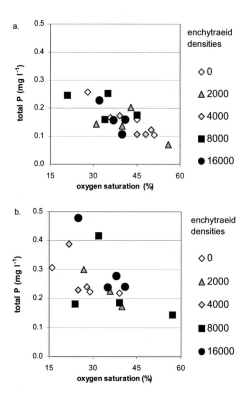

Figure 8.2: Correlation of total P and O$_2$ saturation of the flood water in treatments with different enchytraeid densities a. without *O. cyaneum* (r = -0.7; p < 0.01), b. in the presence of *O. cyaneum* (r = -0.61; p < 0.01, Spearman coefficients)

Unlike in the previous experiment (Chapter 7), there was no correlation between concentrations of ammonium and pH of the water, nor for all treatments together (n = 40), nor for each single treatment (n = 4). Besides, no significant quadratic regression for NH$_4^+$-concentrations or water pH as affected by enchytraeid density was found based on the whole data set. However, a density-dependence of measured parameters was found for the 10 treatments of single replicates, e.g. for the pH of the flood water. The effect varied over time, with a clear pattern being only visible in replicate 2 (Fig. 8.3). Only here the hypothesised pattern was observed whereas in the other series pH rather tended to decrease with increasing enchytraeid density. No such effects were found for any of the nutrients, except for a tendency in nitrate concentration (cf. Fig. 8.1).

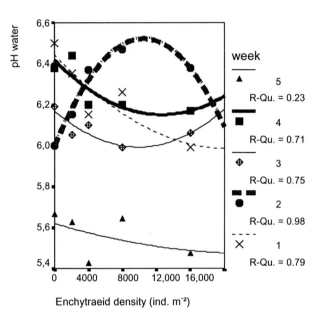

Figure 8.3: Density-dependence of water pH in flooded microcosms in time. Treatments with enchytraeids only. Each line represents the values of one of the 5 replicates carried out in 5 weeks running (week 1 starting on April 13th, week 5 starting on May 10th with soil cores taken in April instead of March) with enchytraeid black box communities taken from fresh soil. R-Qu. = Root-square-values of correlation coefficients calculated by the SPSS statistical programme.

8.3.4 Influence of earthworms

Independent of enchytraeid density, the oxygen saturation in treatments with *O. cyaneum* was significantly lower (T = 2.8, p < 0.01), whilst the concentration of ammonium (T = -3.1, p < 0.001) and total phosphorus (Z = -3.96, p < 0.001) was significantly higher (Fig. 8.1). Besides, treatments with *O. cyaneum* had a lower water pH (Z = -2.3, p < 0.05) and a higher electrical conductivity (Z = -2.0, p < 0.05; n = 20 for all parameters). A two-factorial ANOVA showed about the same results: *O. cyaneum* had an influence on oxygen saturation (F = 9.96, p < 0.01), water pH (F = 5.14, p < 0.05), electrical conductivity (F = 4.7, p < 0.05), concentrations of NH_4^+-N (F = 9.6, p < 0.01) and total phosphorus (F = 11.92, p < 0.01) and no influence of enchytraeid density or a combination of both factors.

In treatments with *O. cyaneum*, electrical conductivity and concentrations of NH_4^+-N and total P correlated positive with each other, while the oxygen saturation correlated negatively with these three parameters as well as with soil pH (Table 8.1). The correlation of oxygen saturation and total P was less pronounced than in treatments with enchytraeids only (Fig. 8.2).

For ammonium-N and water pH, very good quadratic regressions (Rsq = 1.0) were found for the treatments without enchytraeids (p < 0.05) and with the lowest density (p < 0.01; n = 4 for both), but they contrasted: without enchytraeids, a maximum pH would occur with NH_4^+-N values around 18 mg l^{-1}, while in the presence of 2,000 enchytraeids per m-², a minimum pH would occur close to this at 15.8 mg l^{-1} NH_4^+-N.

Table 8.1: Spearman correlation coefficients for treatments with *O. cyaneum*

Parameter	O_2 saturation (%)	electr. conduct. (μS cm^{-1})	NH_4^+-N (mg l^{-1})
electr. conduct. (μS cm^{-1})	-0.61**		
NH_4^+-N (mg l^{-1})	-0.47*	0.68**	
total P (mg l^{-1})	-0.61**	0.57**	0.7**
soil pH	-0.58**		

n = 20 (only soil cores taken in March).
* p < 0.05; ** p < 0.01

8.4 Discussion

8.4.1 Surviving animals

The fact that all earthworms were found alive inside the soil cores hints at a good physiological condition, as they tend to come to the surface in the case of oxygen deficiency. In a previous experiment, even less enchytraeids (about 3%) than here could be found back, corresponding to the longer duration of the simulated inundation (72 h). Also in the field, a high mortality of enchytraeids during the first days of a late autumn inundation at even lower temperatures was recorded (Chapter 5). However, the extraction method used only works efficiently when the worms move and fall down through the sieve into the collecting flask. It is possible that not all enchytraeids died, but that most of them became inactive because of the oxygen deficiency and were simply not extracted, while the decay of their bodies had not yet started. This has to be considered in the interpretation of nutrient release. Besides, an additional loss of enchytraeids in the portion of soil used for the pH determination has to be taken into account. As no nematodes had been found in the animal-free controls, they probably did not hatch from frost-resistant eggs in the defaunated soil cores, but were brought in with the annelid worms.

8.4.2 Mobilisation of nutrients

The general high release of nutrients could be explained by the preparation of the soil by freezing and thawing (Kampichler et al. 1999). The peak in the release of nitrate in the first week of April (5th microcosms series) compared to that in winter or later in spring was also described from field studies on different peat soil sites in the nature reserve "Borgfelder Wümmewiesen" (Brünjes 1994), Rodieck et al. 1992). The phenomenon was explained with the warming up of the soil in spring, implying an increased mineralisation, while plants are not ready for the take-up of released nitrate.

8.4.3 Effects of enchytraeids

In general, density-dependent effects of enchytraeids could hardly be found in this experiment, ands if so they varied over time. The very short incubation time followed by the high mortality and/or the inactivity of enchytraeids when flooded are probably the main reasons for this. Fresh enchytraeid faeces are rich in nutrients (Schrader et al. 1997), but these are occluded by the compact structure of the casts and only leached later on (Wolters 1988). The amount of nutrients from the tissue of dead enchytraeids is very small (Williams and Griffiths 1989) and probably did not have a measurable influence on nutrient concentrations or oxygen saturation of the flood water in the present experiment.

The absence of a correlation between ammonium and pH of the flood water in the presence of enchytraeids like in the previous experiment could be attributed to the difference in temporal design: In the present experiment, measurements were made earlier, i.e. after 24 h instead of 72 h, and a different "snap shot" of the processes involved was taken. Probably, this correlation only occurs when high amounts of ammonium are released from dead tissue together with buffer substances that bind H^+ ions. This is confirmed by the results of another experiment with flooded peat soil cores (Chapter 9) where a positive correlation NH_4^+-N - water pH was found again in the treatment with the highest amount of dead earthworm tissue. Since oxygen is a precondition for nitrification, the oxygen saturation of the water correlates positively with nitrate-N concentrations.

The simulated enchytraeid densities are realistic densities often found in the field. During two years of field investigation, a mean density of 8,100 enchytraeids m^{-2} with a mean biomass of 0.032 g was found on the study site. Maximum abundances of 50,000 ind m^{-2} were recorded in the peat soil in a dry summer period (Chapter 5).

As the enchytraeids were extracted from incubated soil from the field for each of the five replicates, differences in population structure (e.g. portion of juveniles with a more active metabolism than adults) could have influenced the results. Thus, a density-dependent effect like observed in other microarthropods, e.g. Collembola (Finlay 1985) could be valid for enchytraeids in flooded soils, but only under certain seasonal circumstances.

8.4.4 Effects of earthworms

The mobilising effect of *O. cyaneum* on ammonium and phosphorus from peat soil into flood water is reproducible. In a previous experiment, the effect was more pronounced, related to the higher earthworm density (600 instead of 200 ind. m^{-2}) and the longer experimental time (7 days instead of one for earthworms to adapt, 72 h of flooding instead of 24 h; Chapter 7). It is very likely that nutrients are mobilised from the earthworms' casts that are known to be rich in inorganic phosphorus and mineral nitrogen (Haynes et al. 2003, Ruz-Jerez et al. 1992, Scheu 1993, Sharpley and Syers 1977). Also the positive correlation of NH_4^+ and total P hints at a common source that could be the earthworms' casts. The more nutrients (and other substances that were not measured) are released into the water from the earthworms' excretes, the higher the electrical conductivity and the lower the oxygen saturation of the water. This could also cause death and decay of enchytraeids. The same combination of correlations was found in an experiment with living and dead earthworms (Chapter 9) and could be explained by a combination of release from excretes and decay of dead tissue.

The correlation of oxygen saturation with total P could be less pronounced than in treatments with enchytraeids only because the earthworms foster oxygen-consuming and P-limited microorganisms by providing phosphorus for their growth.

8.4.5 Conclusions

The present study corroborates previous findings that flooded peat soils with a large oligochaete biomass have a high mobilisation potential for ammonium and phosphorus. Nutrient mobilisation from flooded soils is a complex, dynamic process. Values measured in microcosm experiments are just snapshots that highly depend on time, i.e. the time (season) the soil cores were taken in the field, the time between thawing and flooding, the duration of flooding and community structure of soil organisms (here: enchytraeids). Additionally, numerous other substances are released from soil and interact with the measured nutrients and proton concentration. Under field conditions, differences of annelid effects can be expected between soils and in time. This should be taken into account in controlled water management.

8.5 Acknowledgements

We are grateful to several members of our department as well as of the Institute of Environmental Process Engineering (IUV) for their practical assistance in the field and laboratory work. We also thank Ulfert Graefe, IfAB Hamburg, for advice in the identification of enchytraeids and members of the Senator für Bau und Umwelt Bremen and the WWF (Worldwide Fund for Nature) for giving permission to take samples on the study site.

9. Dead or alive: The effects of living and dead earthworms on nutrient mobilisation dynamics from a flooded peat soil

submitted to *Biogeochemistry* as: Plum, N.M.: Effects of living and dead earthworms (Lumbricidae) on nutrient mobilisation dynamics from a flooded peat soil (under revision)

Abstract: The microcosm experiments presented in this paper investigate the role of earthworms in the biogeochemical functioning of a river floodplain meadow. A main purpose of wetlands is the reduction of nutrients in river water. From studies on terrestrial systems, the enhancing effect of earthworms on nutrient release from soil and litter into leaching water and surface runoff is well known. The experiments presented here investigate the effect of two earthworm species (*Octolasion cyaneum* and *Lumbricus rubellus*) and of their dead tissue on floodwater chemistry within the first 24 hours of an inundation event.

Flooding events were simulated on undisturbed peat soil cores in microcosms with river water. In the first experiment, concentrations of ammonium-N) and nitrate-N) of the flood water increased constantly and were significantly higher (192% for NH_4^+-N, 217% for NO_3^--N) in the presence of *O. cyaneum* after 24 hours compared to an animal-free control (= 100%). Total phosphorus concentrations in the presence of *O. cyaneum* were lower than in the control treatment after five hours (40% of the control) but tended to be higher again after 24 hours (248%). In the second experiment, the living *L. rubellus* used for the inoculation of microcosms had no significant effect on any of the measured parameters. Compared to the animal-free controls (= 100%), added dead tissue of earthworms led to an increased concentration of ammonium (229%) and total phosphorus (max. 243%) and to a reduction of nitrate concentrations of the water down to 32%, probably due to denitrification but also to an impeded nitrification.

9.1 Introduction

The microcosm experiments presented in this paper investigate the role of earthworms in the biogeochemical functioning of a river floodplain meadow. Such terrestrial-aquatic interfaces are "biogeochemical hot spots", i.e. areas that show disproportionately high reaction rates relative to the surrounding area (McClain et al.

2003). The same authors coined the term "biogeochemical hot moment", i.e. the moment when accumulated reactants are reactivated or mobilized by hydrological flow paths (McClain et al. 2003). A main purpose of controlled flooding of wetlands is the reduction of undesired nutrients and pollutants in the flood water (Haslam et al. 1998). From studies on terrestrial systems, the enhancing effect of earthworms on nutrient concentrations in leaching water and surface runoff after heavy rains is well known. Living earthworms foster the growth of microorganisms involved in nitrification (Scheu 1993, Weiß 1994, Haimi 1995, Helling 1997), while ammonium is released from excretes and dead tissue (Sharpley and Syers 1977, Ruz-Jerez et al. 1992, Haynes et al. 2003).

The main source of the released nutrients is the earthworms' faeces (usually called "casts"). They are known to be enriched in ammonium and phosphorus compared to the surrounding soil; Sharpley and Syers 1977, Parkin and Berry 1994). By their burrowing, the earthworms also release nutrients from soil and litter (Scheu 1993, 1995). Next to this, the tissue of earthworms is an important source of nitrogen that is quickly mineralized after death (Christensen 1987, Curry 1994). As earthworm populations are often reduced to zero during flood events (Ekschmitt 1991, Faber et al. 2000, Ausden et al. 2001, Zorn et al. 2004a, Chapter 5), the role of their dead tissue in nitrogen cycling and phosphorus dynamics may be of great importance under field conditions.

The following hypotheses were tested: (i) The temporal dynamics of nutrient release would not be linear, but fluctuating, especially with regards to ammonium concentrations. (ii) Earthworms would have an increasing effect on concentrations of NH_4^+, NO_3^- and total phosphorus in the floodwater. (iii) The effect of earthworms on nutrient release would not be visible at once (i.e. after an hour), but only after a period of time (i.e. a day). (iv) As during the decay of organic material under water oxygen is consumed and salts are released, it was assumed that in the presence of dead earthworm tissue, the oxygen saturation of the floodwater would be lower and the electrical conductivity higher than in the animal-free control. (v) Ammonium and phosphorus would be mobilized from dead earthworm tissue. (vi) The effect of the epigeic (surface-dwelling) earthworm *Lumbricus rubellus* (Hoffmeister) would be less pronounced than that of the endogeic (soil-dwelling) *Octolasion cyaneum* (Savigny) as the latter moves greater amounts of soil by burrowing.

In two previous microcosm experiments with flooded soils, earthworms enhanced the mobilization of nutrients. Especially *O. cyaneum* in peat soil had a mobilizing effect on ammonium and phosphorus, while in the presence of dead *L. rubellus*, a slightly higher concentration of total phosphorus, but not of ammonium was recorded in the flood water on a marsh and a gley soil (Chapter 7). The aim of the following experiments was to shed light on the influence of earthworms on nutrient release dynamics. At the same time, the different role of different life forms of earthworms and of their dead tissue were investigated.

9.2 Material and Methods

In two laboratory experiments, flooding events on soil cores in microcosms were simulated. In the first, temporal dynamics of nutrient release from a peat soil into floodwater in the presence and absence of the earthworm *O. cyaneum* were measured at different times within the first 24 hours of inundation. The second experiment, using the same peat soil and inundation technique, combined living *L. rubellus* with dead tissue of *L. rubellus* and of *O. cyaneum*. Dead tissue of these two species was added to separate treatments and thus detect differences in nutrient release due to a species-specific chemical composition of the tissue.

The study site is a regularly flooded sedge meadow (Caricetum gracilis; Calthion) on peat soil (10-15 cm peat on river sand, Janhoff 1992) in the nature reserve Borgfelder Wümmewiesen in Bremen. The soil had a pH (CaCl$_2$) of 4.6, a maximum water holding capacity of 67 vol. %, a C/N ratio of 14.8 and a soil organic matter (SOM) content of 27.8 % (B. Schuster, pers. comm.).

For both experiments, undisturbed soil cores (d = 8 cm, depth = 5 cm; 10 for the first, 24 for the second experiment) were collected from the study site. Most of the litter was removed carefully. Then the cores were defaunated by freezing at -18°C and thawing. The thawed soil cores were incubated for 4 days at 15°C before they were put into microcosms. Earthworms collected from the study site were kept in boxes filled with peat soil from their habitat and black alder leaves at 15°C in the laboratory for about 2 weeks. The earthworms were set on wet filter paper for the night prior to being used for the experiment microcosms. The fresh weight of earthworms was determined immediately before they were used to inoculate the microcosms.

For the first experiment, concerning the temporal dynamics of nutrient release (in the following referred to as "dynamics experiment"), ten microcosms were made by enclosing sections of the peat soil cores (upper 5 cm) in steel rings closing tightly around the peat soil cores (127 ± 20 g fresh weight). The rings were placed in plastic lids to prevent water from dropping out. Half of them (n = 5) were inoculated with four adult individuals of *O. cyaneum* each (i.e. 800 ind. m^{-2} with a fresh weight of 0.67 ± 0.13 g per microcosm). Microcosms were flooded with 220 ml of filtered water of the river Wümme the next day (giving a water-to-soil ratio of 1.76 and about the same flood height as observed in the field, i.e. about 5 cm). Oxygen saturation, electrical conductivity, and water pH were measured after 1, 5, and 24 hours (using a WTW Oximeter Oxi340i/Set with a DurOx 325-3 probe, a WTW Cond340i probe, and a Knick 766 Calimetric pH probe). At each time, 10 ml water samples were taken with a pipette for direct analysis of ammonium and nitrate. Samples for total phosphorus were stored in acid-washed glass vessels in the refrigerator and analysed the next day. All nutrient analyses were performed photometrically on a CADAS 100 photometer (with Dr. Lange cuvette tests: hypochlorite-salicylate method for ammonium, dimethylphenol method for nitrate, and molybdeneblue method for phosphorus after break up by cooking in acidic environment; www.lange-hach.de). The amount of water remaining within the microcosms at end of study (i.e. the portion of water that had not been absorbed by the soil core) was determined. The number of living earthworms at the end of the study was determined by hand sorting of the soil cores. A sub-sample of soil was used for the determination of soil pH ($CaCl_2$).

For the second experiment (in the following referred to as "dead tissue experiment"), plastic flasks (d = 8 cm) were filled with undisturbed soil cores (198 ± 0.55 g fresh weight, same height and diameter as in the first experiment, but heavier because of a higher water content as they were sampled on the study site in a more humid period of the year). After one week, four different treatments were established by inoculating the microcosms with (i) living *L. rubellus* only, (ii) living *L. rubellus* combined with dead tissue of *L. rubellus*, (iii) living *L. rubellus* combined with dead tissue of *O. cyaneum*, respectively, or (vi) left as an animal-free control (n = 6 for each treatment). To obtain approximately the same amount of dead earthworm tissue in each microcosm, two or three individuals of different age and size were combined (fresh weight of 0.93 ± 0.02 g for *L. rubellus*, 0.87 ± 0.03 g for *O. cyaneum*) and frozen at -18°C in plastic bags.

Twenty-four hours before flooding, 18 of the 24 microcosms were inoculated with one living individual of *L. rubellus* each. The living *L. rubellus* individuals (one adult per microcosm, i.e. amounting to 200 ind. m^{-2}) had a fresh weight of 0.65 ± 0.11 g. The dead tissue was taken out of the freezer one hour before flooding and added to the microcosms immediately before being flooded with filtered river water (220 ml, giving a water-to-soil ratio of 1.11, and, because of the wetter soil cores absorbing less water, again a flood height of 5 cm). Measurements as described for the first experiment were carried out after 24 hours of flooding.

Statistics

Data were tested for statistical preconditions (Kolmogoroff-Smirnov test of normal distribution, Levene-test for homogeneity of variances). Treatments with and without *O. cyaneum* were compared with a T-Test, otherwise with a Mann-Whitney-U-Test. The influence of time and treatment on water parameters was tested with a repeated measurements ANOVA.

For the second experiment, the influence of the treatment on the measured parameters was tested with a 1-way-ANOVA (followed by a Bonferroni PostHoc-test) or a Kruskal-Wallis-H-Test (followed by a Tamhane PostHoc test). Spearman correlation coefficients were calculated in SPSS, version 12.0, for all variables.

9.3 Results and Discussion

Temporal dynamics with *O. cyaneum*. All earthworms survived the experiment. Casts on the surface of the soil core were recorded in all treatments with earthworms. During the 24 hours of the experiment, a similar decrease in oxygen occurred in both treatments, with and without *O. cyaneum* (Fig. 9.1a). In the end, the oxygen saturation was slightly lower in the presence of earthworms, probably because the earthworms themselves as well as the microorganisms in their faeces consumed oxygen. The conductivity constantly increased in the treatment with earthworms, while it decreased slightly in the control treatment after one and five hours, rising again after 24 hours. The conductivity in the control remained at a lower level than the earthworm treatment throughout (Fig. 9.1a). The pH of the water decreased in all microcosms as the neutral floodwater came in contact with the acidic soil. With a decrease from pH 7 to 5.9 ± 0.24 compared to 5.4 ± 0.05 in the control, the earthworm treatment had a significantly higher water pH (F = 21.7, p < 0.001; Fig.

9.1b). Earthworm casts absorb H^+ ions and their mucus can neutralize acidic as well as alkaline environments (Schrader 1994).

Concentrations of ammonium and nitrate in the floodwater on peat soil increased constantly and were higher in the presence of *O. cyaneum*, but the difference was only significant for ammonium after 24 hours (T = 2.4, p < 0.05, Fig. 9.1c, Fig. 9.2). In one microcosm, one of the four earthworms had died during the experiment, and in spite of the three remaining living specimens, the concentration of ammonium was as low as in the lowest controls (Fig. 9.2). When this replicate was omitted from the statistical analysis, the difference to the control treatments was even more pronounced (T = 4.2, p < 0.01). This result evoked the idea to perform an experiment with dead earthworm tissue to determine its role in nutrient release.

In two previous experiments, *O. cyaneum* had already exhibited an ammonium-mobilizing effect. Earthworm excretions (faeces, mucus) were considered as a possible ammonium source, (Chapter 7). The earthworms had emptied their guts on filter paper in the night preceding the inoculation of microcosms. The uptake of food and the passage through the gut take 6 to 8 h (Daniel and Anderson 1992). Thus, after 24 hours inoculation time, the presence of fresh casts was expected and subsequently observed. Binet and Le Bayon (1999) reported a much shorter lifetime of fresh casts under wet conditions (4 days) than during a dryer period (up to 14 days). However, in their study casts did not disappear at the first rain event but only during the second and third rain event. In the present experiment, this could explain the retarded release of ammonium in spite of the presence of casts at the beginning of flooding. In all studies about nutrient release induced by earthworms (Sharpley and Syers 1977, Ruz-Jerez et al. 1992, Scheu 1993, Weiß 1994, Haimi 1995, Helling 1997, Haynes et al. 2003), nutrients are transported from above to below (by percolating or runoff water). A fresh aspect of the present study is the opposite direction of nutrient displacement in stagnant water, from below to above.

The earthworms stayed alive and active during the inundation, so additional casts could have led to a further accumulation of ammonium. The constantly low ammonium concentration over 24 hours in the replicate where one *O. cyaneum* had died cannot be explained by the absence of its faeces alone. There were three specimens left that kept on casting. Thus, microorganisms growing on the dead tissue may have immobilized nitrogen. However, NH_4^+-N concentration in presence of dead earthworm tissue were higher in the second experiment (see below). The replicate in question also deviated for other variables from the mean values of the other four replicates (a lower nitrate concentration, a lower conductivity and a lower oxygen saturation and water pH after 24 hours). An anomaly in the soil core (e.g., a particular high content of clay absorbing H^+- ions like ammonium and thus reducing the electrical conductivity in the water) could have played a role.

For total phosphorus, the temporal dynamics were more complex, as its concentration in the control treatments increased to a maximum value after five hours, displaying a subsequent decline. In the presence of *O. cyaneum*, the concentration after five hours was as low as after one hour, giving a significant difference to the control treatments (U = 2.0, p < 0.05). However, the end-concentration after 24 hours was again higher than in the control (not significant, Fig. 9.1d). Phosphorus limits the growth of microorganisms living in the earthworms' guts and is immediately used by them to reproduce (Karsten 1997). Therefore, it is very likely that microorganisms temporarily immobilize phosphorus. Later, when casts are mineralized and organic phosphorus is turned into the inorganic form (Satchell and Martin 1984, Scheu 1987), concentrations in the water increase again.

Time had a significant influence on all measured parameters, while the treatment only had an influence on water pH, conductivity, and ammonium concentrations of the floodwater. Time * treatment had an influence on water pH, conductivity, ammonium-N and total phosphorus concentrations (RM-ANOVA, Table 9.1).

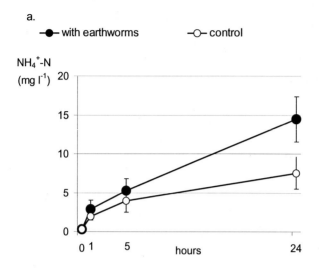

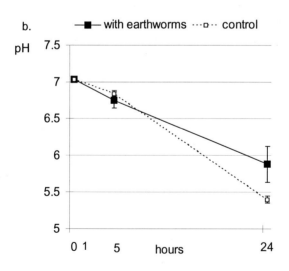

Figure 9.1: Development of a. oxygen saturation and electrical conductivity, b. pH of the flood water in soil microcosms with and without the earthworm *Octolasion cyaneum* during the 24 hours of the „dynamics experiment". Mean values ± SD, n = 5

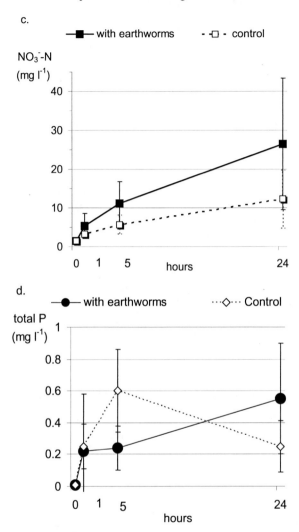

Figure 9.2 (continued) : Development of c. nitrate concentration and d. total phosphorus concentration of the flood water in soil microcosms with and without the earthworm *Octolasion cyaneum* during the 24 hours of the „dynamics experiment". Mean values ± SD, n = 5

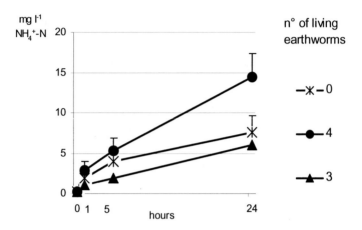

Figure 9.3: Development of ammonium concentration of the flood water in soil microcosms during the 24 hours of the „dynamics experiment". 0 earthworms = control treatment.

Table 9.1: Influence of the factors "time" and "treatment" on water parameters

Parameter	time		treatment		time * treatment	
	F =	p <	F =	p <	F =	p <
pH water	512.8	0.001	21.7	0.001	18.7	0.01
conductivity	10.3	0.01	8.9	0.05	4.6	0.05
O_2 saturation	328.2	0.001	n.s.		n.s.	
NH_4^+-N	66.3	0.001	6.3	0.05	5.3	0.05
NO_3^--N	9.9	0.01	n.s.		n.s.	
total phosphorus	12.0	0.001	n.s.		6.1	0.01

Results of repeated measurements ANOVA with the between-subject-factor "treatment". "Time": measurements in the beginning of the experiment (river water), after one, five and 24 hours of flooding. "Treatment": microcosms with and without *O. cyaneum*.

Effects of dead tissue and living *L. rubellus*. Out of the 18 living *L. rubellus* used for the inoculation of microcosms in the second experiment, 15 survived (two dead specimens were found in microcosms with dead *L. rubellus*, one in those with dead *O. cyaneum*). As compared to plain river water, ammonium and phosphorus were released, the electrical conductivity rose, while pH and oxygen saturation decreased in the floodwater of all microcosms, including the earthworm-free control (Fig. 9.3). The treatment (presence or absence of [dead] earthworms) had an effect on electrical conductivity ($F = 3.7$, $p < 0.05$), oxygen saturation ($F = 20.7$), ammonium ($F = 28.9$) and total phosphorus concentration ($F = 23.9$; all ANOVAS with $p < 0.001$) as well as on nitrate concentrations (Kruskal-Wallis $X^2 = 7.9$, $p < 0.05$) of the floodwater.

The density simulated with one living *L. rubellus* (i.e. 200 ind. m^{-2}) was obviously too low to show any effect on any of the measured parameters compared to the control. In a previous experiment, more ammonium, phosphorus, and nitrite were mobilized from a marsh and a gley soil in the presence of three *L. rubellus* per microcosm (i.e. 600 ind. m^{-2}) compared to a worm-free control (Chapter 7). In another experiment, even a single *O. cyaneum* (i.e. 200 ind. m^{-2}) had a significant effect on mobilization of ammonium and phosphorus from the same peat soil as used for the present experiments (Chapter 8). This corroborates the hypothesis that because of its burrowing (adding further nutrients to those released from excrements), the effect of the soil-dwelling *O. cyaneum* would be more pronounced than that of the surface-dwelling *L. rubellus*.

Both treatments with tissue of dead earthworms differed significantly from the control treatment as well as from the treatment with only one living *L. rubellus*: they showed a lower oxygen saturation of the water ($p < 0.001$) and a higher concentration of ammonium and total phosphorus (for both parameters: $p < 0.01$ for LL, $p < 0.001$ for LO; Bonferroni test, Fig. 9.3). Nitrate concentrations were lower than in the river water used for flooding in these two treatments, while they were higher in the control and in the L treatment (Fig. 9.3e).

Christensen (1987) reported that earthworm tissue was completely mineralized within three weeks at a temperature of 12°C. After one month, 75% of the N previously stored in the earthworm tissue were found back in soil. However, it is possible that the freezing and thawing of the earthworms in experiment 2 caused a faster mineralization of their tissue than after the natural death of the specimen in the first experiment (Mack 1963).

For all treatments together ($n = 24$), the electrical conductivity and the concentrations of ammonium and total phosphorus correlated positively with each other. All three parameters correlated negatively with concentrations of nitrate and with oxygen saturation, while these two parameters correlated positively with each other. When significant correlations were found in single treatments (n = 6), they proved the general correlations, with the exception of the treatment LL, where oxygen saturation and ammonium concentration correlated highly positively, just as ammonium and water pH (Table 9.2, Fig. 9.4). These correlations as well as the lower oxygen saturation and the higher electrical conductivity in treatments with dead earthworm tissue reflect the processes of decay in the water, indicating that earthworm tissue was mineralized and that salts, ammonium, and phosphorus were released. Consequently, the electrical conductivity increased. At the same time, the microorganisms involved in mineralization consumed oxygen, resulting in reduced nitrification. This could also explain the lower nitrate concentrations in the treatments with dead earthworm tissue as well as the correlations shown in Fig. 9.4. The micro-anaerobic conditions may impede nitrification or foster denitrification (Well et al. 2002, Hefting 2003), thereby causing the lower nitrate concentrations observed in the presence of dead earthworms. A third process that plays an important role in stagnant water is the dissimilatory nitrate reduction (or nitrate ammonification; Atlas and Bartha 1998), in which facultative anaerobic bacteria reduce nitrite (that may be excreted by earthworms or released from their dead tissue) via hydroxylamine into ammonium. The negative surfaces of ammonifying microorganisms growing on tissue and casts absorb H+ ions, and the pH rises together with the concentration of ammonium released from the same material. Such a trend was observed in the treatment with living and dead *L. rubellus*.

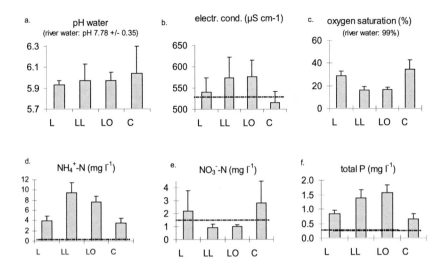

Figure 9.3: Water parameters in flooded microcosms with the treatments L = living *L. rubellus*, LL = living *L. rubellus* and its dead tissue, LO = living *L. rubellus* and dead tissue of *O. cyaneum*, C = earthworm-free control. Dotted line = values measured in plain river water used for flooding.

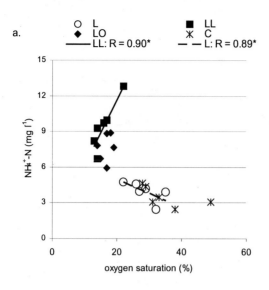

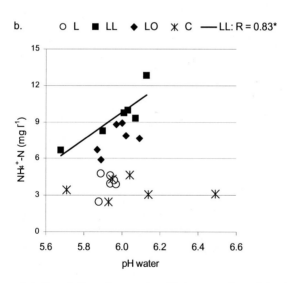

Figure 9.4: Correlation of ammonium-N concentrations of the flood water with a. oxygen saturation and b. water pH in different treatments (see Fig. 9.3).

Table 9.2: Correlations of flood water parameters in the dead tissue experiment

	O$_2$ satur.	electr. conduct.	water pH	NH$_4^+$-N	NO$_3^-$-N
O$_2$ saturation	-				
electrical conductivity	-0.52 **	-			
pH water			-		
NH$_4^+$-N	-0.81** L: - 0.89* LL: 0.90*	0.70** LL: 0.83*	LL: 0.83*	-	
NO$_3^-$-N	0.67**	-0.73** C: -0.94**		-0.68**	-
total phosphorus	-0.80** C: - 0.89*	0.72** C: 0.83*		0.83** LO: 0.89*	-0.64**

Spearman coefficients; n = 24 for all treatments together. For single treatments n = 6 (L = living *L. rubellus* , LL = living and dead *L. rubellus*, LO = living *L. rubellus* and dead *O. cyaneum*, C = control). Significance levels: * p < 0.05, ** p < 0.01.

9.4 Relevance for field conditions

The highest density recorded during field investigations on the peat soil site was 100 ind. m^{-2} for *L. rubellus* or *O. cyaneum* alone and 140 ind. m^{-2} for all earthworms (Chapter 5). However, this low density was due to repeated extended inundations in the two years of study. Rodieck et al. (1992) found a somewhat higher density (176 ind. m^{-2}) in the same reserve. In other floodplain sites, much higher densities occurred and an effect of the earthworms can be assumed (Zorn 2004, Chapter 5).

In microcosms, there is a lack of movement of the floodwater. In the field, there is wind and water flow with changing water table, ensuring a better aeration. Next to this, the moving water at the beginning of an inundation could translocate nutrients to higher and calmer zones of the wetland, while they remain stationary in the laboratory microcosms. However, also in the field, once the flood reaches a certain level, water remains stagnant, often for weeks, before the draining off (discharge) causes water movement, leading to higher oxygen saturation. This can influence microbial processes such as nitrification in the water. Besides, due to the repeated extraction of water samples in the first experiment, the soil-to-water-ratio continuously increased and partly caused the rising concentrations (220 ml flood water minus 2 x 10 ml, i.e. 9%; 200 ml remaining water for the sampling after 24 hours).

In reed beds constructed for wastewater treatment, earthworms clean the gravel substrate by translocating clogging material from the interstices to the surface, thus prolonging the lifetime of the reed bed (Davison et al. 2005). However, the inoculation of reed beds with earthworms to counteract clogging could be counterproductive because of the nutrient-mobilizing effect of the animals.

To conclude, the previously found mobilizing effect of *O. cyaneum* on ammonium and phosphorus is corroborated in the present study. This effect only becomes evident after 24 hours; thus, its role in nutrient mobilization during very short inundations, as typically occurs in tide-influenced floodplains, may be negligible. High amounts of ammonium and phosphorus are mobilized from the tissue of dead earthworms. Thus, earthworm mortality might be a key regulator of the water quality impacts during flood events. It requires further study to determine if the simultaneous decrease of nitrate concentrations is due to enhanced denitrification or reduced nitrification.

An important purpose of wetlands is the reduction of nutrient concentrations in the river water. Therefore, controlled flooding of wetlands should aim at impeding the death of wetland-dwelling earthworms and at stabilising their populations. This can be achieved when the soil is allowed to drain in spring, then it can be inundated again during the breeding season when earthworms are needed as food resource for meadow birds (Ausden et al. 2001). An artificially raised water table in summer, minimising negative effects of summer drought, should also be favourable for earthworm populations (Ekschmitt 1991, Chapter 5), thus ensuring the food resource for the next bird generation, and minimising undesired nutrient release into the river.

9.5 Acknowledgements

I am grateful to Juliane Filser for helpful remarks on the manuscript. I also thank several colleagues from my department and from the Institute of Environmental Process Engineering (IUV) at the Center for Environmental Research and

Technology (UFT), Bremen, for their practical assistance in field and laboratory work as well as members of the Senator für Bau und Umwelt Bremen and the WWF (Worldwide Fund for Nature) for giving permission to take samples on the study site. My special thanks go to Tom Headley (Southern Cross University, Australia) for linguistic correction.

10. A field experiment with earthworms in flooded mesocosms

Abstract: A field experiment with flooded soil mesocosms, inoculated with *L. rubellus*, was set up in September 2004 in the marsh site. The aim was to assess if the results obtained from the laboratory experiments described in Chapters 7 to 9 are also valuable under natural conditions, i.e. if nutrient concentrations in the flood water are increased after contact with the soil, in spite of plant growth, and if the nutrient-mobilising effect of earthworms could be found back.

The experiment failed as the river water used for flood simulation trickled away within some minutes in half of the mesocosms. In the other half, nutrient concentrations were determined after 1.5 hours of flooding. In spite of the short flooding time, concentrations of nitrate-N, ammonium-N and total P in mesocosm water were remarkably higher than in the river water used for the flood simulation. For nitrate-N, concentrations were even higher than after 72 hours of inundation in the laboratory. Concentrations of nitrate and ammonium tended to be higher in mesocosms with an increased population of earthworms compared to the control with decreased earthworm populations. However, differences between treatments were not significant because of the low number of remaining replicates. The general nutrient release is discussed regarding processes within nitrogen cycling and phosphorus transformation. The electrical octett method is regarded as suited for the reduction of earthworm populations on mesocosm plots in the field. The technical problems in installing terrestrial-aquatic mesocosms are discussed. Reasons for the rapid and complete infiltration of the water (soil drought, inclination of soil surface, vole burrows) and possible improvements of the experiment (deeper mesocosms, choice of humid period) are discussed.

10.1 Introduction

Most research on ecosystem processes is performed in laboratory experiments. The main advantages are a controlled environment, speed and statistical power (Kampichler et al. 2001). Field experiments normally require more time and expense. However, they can overcome the simplicity and deficiencies of laboratory based experimental designs and provide a higher degree of realism (Kampichler et al. 1999). In the present study, four laboratory experiments with flooded soil microcosms had shown a predominantly positive effect of earthworms on nutrient mobilisation from three different soils (cf. Chapters 7 to 9). For these experiments, the undisturbed soil cores had been taken from the field and defaunated by repeated freezing. The most important difference to field conditions is the absence of plant growth, as the experimental time was always short (max. 15 days including thawing of the soil cores and adaptation time for annelids). In the few field experiments examining leachate (Weiß 1994) or surface runoff (Sharpley et al. 1979), a predominantly mobilising effect of earthworms on nutrients was found, just as in the numerous laboratory experiments (Scheu 1993, Haynes et al. 2003 and many others).

The aim of the field experiment was to assess if the results obtained from the laboratory are also valuable under natural conditions. As plants take up available nutrients during growth and contribute to the sedimentation of released particles, the following hypotheses were raised:

(i)   During a river flood in a marsh soil, ammonium, nitrate and phosphorus are mobilised from the soil into the water.

(ii)   However, the increase of nutrient concentration in the flood water is less pronounced than in the laboratory.

(iii)   The measurable effect of earthworms on nutrient release is less pronounced than in the laboratory.

10.2 Material and methods

A field experiment with terrestrial-aquatic mesocosms was designed. As it would require very much effort to simulate natural abundances of enchytraeids at the scale of a field experiment, it was restricted to earthworms alone. Laboratory experiments were focussed on the peat soil (cf. Chapters 3 and 5) which showed the most pronounced mobilisation effect of earthworms. Thus, the same study site was chosen for the field experiment. The peat site was searched for individuals of earthworms of *O. cyaneum*, the dominating species that already had been used for the inoculation of laboratory microcosms. As in the course of several months (June to September 2004, four sampling dates), only few earthworm individuals were found, I had to switch to the marsh site ("Ochtumniederung bei Brokhuchting", cf. Chapter 3).

The dominating earthworm *L. rubellus* was collected on the study site in sufficient numbers and held in the laboratory until the start of the experiment for four days. Spots for the installation of mesocosms were defaunated with the electrical octet method (Thielemann 1986). Appearing earthworms were collected. The *L. rubellus* amongst them were added to the laboratory culture, other species were brought to a remote corner of the study site.

Earthworms were held on wet filter paper the night preceding the inoculation of microcosms. In the morning, their weight (groups of 5 individuals, 3.05 ± 0.39 g per microcosm) was determined. The groups were transported to the study site in separate containers.

Mesocosms were prepared by removing the bottom of plastic buckets, giving broad plastic "rings" (upper $\emptyset$ = 24 cm, lower $\emptyset$ = 20 cm, 25 cm high). Out of twenty possible spots, those for the fourteen mesocosms (7 with earthworms, 7 as earthworm-free control) were determined randomly using dices.

Mesocosms were embedded into the soil in a depth of 5 cm and half of them (ns° 1 to 7) inoculated with earthworms. With five *L. rubellus* individuals per mesocosm, an abundance of 110 ind. m^{-2} was simulated. This is within the range of the mean earthworm abundance recorded during two years of field investigations (87 ± 60 ind. m^{-2}). They were given three days to adapt to the new environment. During this time, minimum and maximum temperature (-4°C at night up to +20°C at noon) as well as precipitation (no precipitation during the experiment) were measured. Then the mesocosms were flooded with 2.1 l of filtered water taken directly from the river Ochtum, giving the same relation of water to soil as in the laboratory experiment (vol:fw = 1.6). This first trial, conducted on the "slopes" of the water-filled ditches, completely failed as overnight the water trickled away from all mesocosms.

When the second trial was started on October 8, 2004, there was no stagnant water in the shallow ditches and mesocosms could be placed directly where the ground water table was closest to the surface (Fig. 10.1). When the mesocosms were flooded on October 11, it was planned to take a first set of water samples after 1.5 hours of flooding and if possible, further samples after 24 and 72 hours (the measuring time of the corresponding laboratory experiment). However, the water only remained in half of the mesocosms (bold circles in Fig. 10.1). In these, water samples were taken after 1.5 hours. The other mesocosms were flooded again with an additional liter of filtered river water, and water samples were taken immediately after two minutes. Ammonium, nitrate and total phosphorus were determined photometrically as described in Chapter 8.

After water samples had been taken, all mesocosms were filled up with 3 l of unfiltered river water to observe the further development. In four of the mesocosms

(Fig. 10.1, bold circles), at least half of the water was still there on the third day, indicating a flowing direction of water towards the ditch in the west of the site.

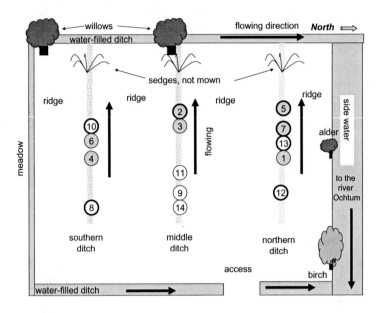

Figure 10.1: Schematic position of mesocosms on the marsh site "Ochtumniederung bei Brokhuchting" (2nd trial of the experiment). Numbered circles = mesocosms; 1 - 7 with *L. rubellus* (grey), 8 - 14 control (white). Bold circles: those where the water remained for 1.5 hours.

10.3 Results

In all treatments (including those filled up and measured after two minutes), concentrations of nitrate-N, ammonium-N and total P were much higher than in the river water used for the flood simulation. Table 10.1 shows the factors of increase after 1.5 hours of inundation in the field compared to those after 72 hours of inundation in laboratory microcosms (Chapter 7). In all treatments of both experiments, this factor was highest for ammonium. In the presence of *L. rubellus*, there was a tendency towards higher nitrate and ammonium concentrations in those mesocosms where the water remained (Fig. 10.2). For phosphorus, there was a trend towards lower concentrations with *L. rubellus*. However, because of the few remaining replicates and the very large standard deviations, the results were not significant. In the filled up mesocosms, nutrient concentrations were distinctly lower than in those where the water had remained, and no differences between treatments were found.

Table 10.1: Mobilisation of nutrients from marsh soil to river water

Mobilisation after...	NO$_3^-$-N		NH$_4^+$-N		total P	
	earthw.	contr.	earthw.	contr.	earthw.	contr.
...1.5 hours in the field	2.9	2.4	25	21	4.3	4.4
...72 hours in the laboratory[a]	1.6	1.8	100	47	100	80

Values indicate the factor of increase of nutrient concentration (i.e. between 2.9 and 100fold) when river water and mesocosm water are compared. "earthw." = treatment with increased earthworm abundance, field: n = 3; laboratory: n = 5; "contr." = control treatment with reduced earthworm populations (field, n = 4) or without earthworms (laboratory, n = 5).

[a] Results from the "three soils experiment" (Chapter 7)

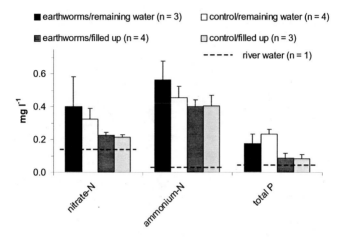

Figure 10.2: Water parameters of field mesocosms. Earthworms: with 5 *L. rubellus*; control: without inoculation of earthworms; natural population reduced by electrical octett method. Water remaining: Mesocosms where the water remained and samplings could be taken after 1.5 hours; filled up: Mesocosms where the water trickled away immediately; they were filled up with another liter of river water, then samples were taken after two minutes.

10.4 Discussion

10.4.1 Nutrient release

The mobilisation of nutrients from all treatments, irrespective the short duration of flooding, is remarkable. Conservation of wetlands normally aims at reducing nutrient concentrations in the river water (Haslam 2003). Obviously, the growth of plants in the field makes no difference for this process: The amount of nutrients released from soil is greater as the amount taken up by plants or brought to sedimentation by them. However, as plants were cut short also in this field experiment, their impact on sedimentation could also have been reduced.

The short duration of flooding in the field experiment implies lower concentrations of ammonium and phosphorus than in the 72h-laboratory experiment. However, the increase of nitrate concentrations was greater after 1.5 h in the field than after 72 hours in the lab. It is possible that released nitrate is reduced to ammonium when the inundation continues (re-ammonification; Bowden 1987). On the same site, a very low nitrification rate was measured by Erber (1998) and explained with the frequent inundation by stagnant water (lack of oxygen). However, she could measure nitrate at most sampling dates. This might be a result of the water flow and translocation of solved ions from the dryer ridges.

In a riparian wetland in Denmark nitrate from the river water was reduced during very short inundations (some minutes up to hours). When the inundation lasted for several days, nitrate concentrations in the water leaving the meadow were further reduced down to trace amounts. Occasionally, they increased again to the river water value (Hoffmann et al. 2003). An important difference is the much higher concentration of nitrate-N in the river Gjern studied by these authors (2.4 mg l^{-1} in November; 5.0 mg l^{-1} in January and 5.7 mg l^{-1} in March) compared to the river Ochtum (0.14 mg l^{-1} in September '04; 1.3 mg l^{-1} in July '03). It is very likely that high nitrate concentrations are reduced by the flooded grassland, while low concentrations are slightly increased when the water comes into contact with the soil. The season could also play a role, though plant growth is likely to be more pronounced in summer than in the winter months studied by Hoffmann et al. (2003). However, leaching losses can also be reduced by microbes that rapidly colonise dying (plant) tissues. Fresh, aerobic litter of plants may act like a protecting cap that prevents nitrogen loss to the river water, also during winter when there is no plant growth (Bowden 1987). In the present field experiment and also in the laboratory, plant litter was removed and plants were cut short before the inundation of mesocosms. Thus, the protecting effect of litter was absent here and could be a further reason for the increased nutrient mobilisation, just as the movement of soil in installing mesocosms.

10.4.2 Effects of earthworms

The trend towards higher ammonium and nitrate concentrations in the presence of *L. rubellus* corresponds with results from the laboratory: After 72 hours of inundation of marsh soil cores, concentrations of ammonium, nitrate and total phosphorus tended to be higher in the presence of a natural mean density of *L. rubellus* compared to an animal-free control. However, the results were only significant for nitrate and total phosphorus when enchytraeids were present in the same microcosm as the earthworms (Fig. 7.1a). The probable source for these nutrients were the worms' excrements.

In the field, phosphorus concentrations rather tended to be lower in the presence of earthworms. Also in the temporal dynamics experiment with peat soil (cf. Chapter 9), total P concentrations in the presence of the earthworm *O. cyaneum* were only higher than in the control treatment after 24 hours, while after five hours, they were significantly lower. This was explained by a transitional immobilisation in the guts and could also be the case in this field experiment. Sharpley et al. (1979) explained the reduced particulate P transport in surface runoff from a pasture as follows: Earthworms incorporate litter into the soil, thus overcompensating the amount of P leaching from their casts. However, in the present study, most litter was removed from the mesocosms. It is very likely that during an inundation phosphorus and ammonium are only mobilised from casts in measurable amounts when these have accumulated over 24 hours or more (cf. Chapter 9). Thus, a field experiment should at least have this duration.

As only few replicates with *L. rubellus* remained and the results were not significant, no interpretation is possible without further experiments.

10.4.3 Electrical octet method

The electrical octet method covers a sufficient area to reduce earthworm populations on plots for mesocosm field experiments and it is non-destructive (Schmidt and Curry, unpubl.). For this purpose, it has been used successfully by Schindler-Wessels et al. (1997), Blair et al. (1997) and Subler et al. (1998). In experiments about nutrient dynamics, destructive methods, e.g. with chemicals, can have a direct influence on the measured results. Sharpley et al. (1979) used carbaryl to reduce earthworm populations. The consequences were an increased breakdown of organic material and increased concentrations of NH_4^+-N and NO_3^--N in surface runoff, independent of an earthworm effect. Defaunation of soil cores by freezing and thawing, like carried out for laboratory experiments, has a comparable effect (Kampichler et al. 1999).

Bohlen et al. (1995) showed that electro shocking was effective at reducing earthworm populations, expelling about 75% of their abundance. On sites neighboured to the study site in the present experiment, abundances of 15 to 35% of those sampled by hand-sorting were achieved with the same apparatus (Blanken-Mittendorf 1990). The low efficiency was explained by a high electrical resistance of dry soil. The method works better with humid soil conditions (pF 2.0 to 3.2). Only when the soil is too wet, problems with the functioning of the electrical device occur. During the present study, conditions can be considered to be ideal for the electrical method. According to Blanken-Mittendorf (1990), the best results are obtained for epigeic species like *L. rubellus*. However, no control of the efficiency on the same plot was carried out by him, just as in the present experiment where hand sorting would have destroyed the plots for mesocosms. The *L. rubellus* expelled by electro-shocking throve well in laboratory culture. Thus, the method did not seem to damage their health.

10.4.4 Terrestrial-aquatic mesocosms

The simulation of a terrestrial-aquatic environment in field mesocosms revealed some difficulties. A first idea was to use mesocosms closed at the bottom, just like in the laboratory. Undisturbed soil samples would be put into plastic containers inserted into the soil, but emerging high enough to contain about 10 cm of flood water. However, in such a system the natural contact of groundwater to soil and flood water is not given.

In a preliminary experiment with four mesocosms (open-bottom-type like described above) in the peat site (with a high groundwater table), water remained overnight. The complete water remained in those installed in a shallow ditch. In the others, only half of the water was left. This shows that at least a short-term field experiment with open-bottom mesocosms should be possible at comparable conditions. The experiment could not be carried out in that site because no earthworms were found for the inoculation. However, some criteria for an improvement of the field experiment design can be derived: To prevent the flood water from running away, soil depressions like the deepest points of the shallow ditches on the marsh site should be chosen as plots for the mesocosms. Furthermore, the natural flow direction of surface water should be taken into account to determine the ideal position of mesocosms. Fig. 10.1 clearly shows that on the study site, water discharges to the north and to the west. This is why mesocosms with remaining water (bold circles) concentrate in the northern ditch and in the western ends of the ditches.

Another way to improve the experimental setup is to insert mesocosms deeper into the soil to reach the groundwater table, even when it is lower. This would be possible with larger rings, e.g. pieces of broad plastic tubes with a length of at least 1 m. These tubes should be inserted into the soil with mechanical help, i.e. with an excavator or a tractor with front loader. However, the only way to carry out the experiment successfully is to choose a soil-humid period or to raise the water level artificially (if controlled flooding is possible). This again could lead to a mixture of flood water with groundwater and a reduction of nutrient concentrations measured in the flood water. Nevertheless it would reflect natural conditions.

Furthermore, another problem occurs: A dense network of vole burrows was found in the marsh soil, partly visible on the surface (tracks of collapsed grass turf all over the study site). In choosing plots for mesocosms, obvious burrows near the surface were avoided. However, the flood water could have trickled away by preferential flow through subterranean burrows. For laboratory microcosms, undisturbed soil samples were used. A homogenisation of the soil in the field, destroying the vole burrows, would create a bias from natural as well as from laboratory conditions.

The experiment shows that wetlands do not always reduce nutrients in the flood water. From the marsh soil site, nutrients are mobilised, even during inundations of a few minutes.

Concerning the earthworm effects, the results could probably be reproduced for field conditions if the experimental design was improved. A period of humid soil conditions and a vole-free area to install improved (deeper) mesocosms should be chosen for a new experiment.

11. Synthesis

To approach the research question about the role of annelids in the mobilisation of soil nutrients during flooding events, other influences on nutrient dynamics in wetlands in general and especially in the experimental situation have to be analysed and excluded. In section 11.1, the influences of site, soil, chemical composition of floodwater, flooding duration, season, experimental methods, and soil fauna in general are analysed. In 11.2, the different effects of annelids on nutrient dynamics are discussed. The chapter is subdivided in sections about the sources and sinks of nutrients found in the floodwater.

Table 11.1 gives an overview of the experiments carried out in the framework of this thesis. The results are summarised as mobilisation factors in different treatments in Table 11.2. The nutrient concentrations of the microcosm floodwater are divided by the concentration of the river water used for flood simulation to calculate these mobilisation factors.

Microorganisms in the soil cores are reduced by freezing, but can be assumed to be present and active during the experiments. Anaerobic conditions occurred in the field as well as during laboratory experiments. Thus, all the various possible processes of nutrient cycling (cf. Chapter 2, Figure 1) have to be taken into account when results concerning nutrient exchange processes and the role of annelid worms in them are interpreted.

Table 11.1: Experiments carried out for this thesis

Experiment[a]	Chapter	Title	Organisms	Soils
"3 soils"	7	Earthworms and potworms (Lumbricidae, Enchytraeidae) mobilise soil nutrients from flooded grassland	*L. rubellus*, *O. cyaneum*, enchytraeids	marsh, peat, gley soil
"Density"	8	Temporal variation in nutrient release from a flooded peat soil depending on earthworms and enchytraeid density	enchytraeids, *O. cyaneum*	peat soil
"Dynamics"	9	The effect of living and dead earthworms (Lumbricidae) on nutrient mobilisation from a flooded peat soil	*O. cyaneum*	peat soil
"Dead tissue"	9		*L. rubellus*, *O. cyaneum*	peat soil
"Field"	10	A field experiment with earthworms in flooded mesocosms	*L. rubellus*	marsh soil

[a] short titles used in the following text

11.1 Factors influencing nutrient dynamics in terrestrial-aquatic microcosms

Table 11.2 shows that during the flood simulation in terrestrial-aquatic microcosms, ammonium and phosphorus were normally mobilised in considerable amounts from all three soils. Nitrate was generally released to a lesser extent. Immobilisation (in Table 11.2 represented by factors < 1) occurred rarely: twice for nitrate (when soil cores were taken in March and in the presence of dead earthworm tissue) and once for phosphorus (in an animal-free control in the density experiment; all observations for peat soil).

Table 11.3 gives an overview on nutrient concentrations in the river water used for flooding of the microcosms. The order of magnitude of these values and their seasonal dynamics should be kept in mind when the factors of mobilisation shown in Table 11.2 are interpreted in the following chapter.

Table 11.2: Mobilisation of nutrients from soil to river water. Values indicate the factor of increase of nutrient concentration when river water and floodwater in microcosms are compared (e.g. a hundred-fold increase of ammonium-N and total phosphorus, but only a 1.6-fold increase of nitrate-N in the first line). Ch. = Chapter. "L." = treatment with earthworms, *L. rub.* = *Lumbricus rubellus, O. cyan.* = *Octolasion cyaneum*, "E" = treatment with enchytraeids; densities corresponding to ind. m^{-2} are indicated in parentheses; control = treatment without annelids (decreased earthworm abundance in the field experiment). "Month" is the month in which soil cores were taken from the study sites. Values for treatments with different enchytraeid densities were pooled because there were no significant differences. The mobilisation factors show different patterns between sites (cf. 11.1.1) and throughout the seasons (cf. 11.1.3). More nutrients are released with a longer flooding duration (cf. 11.1.2) and normally also in the presence of annelids (cf. 11.2, Table 11.6).

Experiment	Ch.	Dura-tion	Month, Year	Treatment (ind. m^{-2})	n	NH_4^+-N	NO_3^--N	total P
„3 soils", marsh soil, laboratory	7	72 h	May 2003	*L. rub.* (600)	5	100	1.6	100
				E	5	44	2.0	56
				L & E	5	146	3.0	114
				Control	5	47	1.8	80
„3 soils", peat soil, laboratory	7	72 h	May 2003	*O. cyan.* (600)	5	387	3.0	118
				E	5	241	6.6	68
				O. cyan. & E	5	249	3.0	143
				Control	5	136	3.0	50
„3 soils", gley soil, laboratory	7	72 h	May 2003	*L. rub.* (600)	5	12	104	410
				E	5	25	103	301
				L. rub. & E	5	17	90	309
				Control	5	4.4	85	384
"density", peat soil, laboratory	8	24 h	March 2004	*O. cyan.* (200)	4	524	0.5	13
				E [a]	16	411	1.7	9
				Control	4	375	0.6	6
			April 2004	*O. cyan.* (200)	1	24	7.0	13
				E [a]	4	8	5.0	11
				Control	1	17	2.3	7

Table 11.2: Mobilisation of nutrients from soil to river water (continued)

Experiment	Ch.	Dura-tion	Month Year	Treatment (ind. m^{-2})	n	Mobilisation factor NH_4^+-N	NO_3^--N	total P
"dynamics ", peat soil, laboratory	9	1 h	Jan. 2004	*O. cyan.* (800) [b]	4	10	52	20
		5 h			4	19	7.7	21
		24 h			4	52	18	48
		1 h		Control	5	7	2.2	19
		5 h			5	14	4.0	46
		24 h			5	27	8.0	19
"dead tissue", peat soil, laboratory	9	24 h	August 2004	*L. rub.* (200)	6	29	1.7	2.8
				L (dead tissue)[c]	6	62	0.7	5.0
				Control	6	26	2.1	2.2
"field", marsh soil	10	1.5 h	Sept. 2004	*L. rub.* (110)	3	25	2.9	4.3
		1.5 h		Control	4	21	2.4	4.4

[a] all densities pooled

[b] without 5th replicate where one earthworm had died and nutrient concentrations differed drastically (cf. Chapter 9)

[c] means of both dead tissue treatments (with dead *L. rubellus* and dead *O. cyaneum*)

Table 11.3: Nutrient concentrations in water from different rivers

Own measurements. Month: when the water sample was taken. differences between sites (sampling in May 2003) are discussed in 11.1.1, the seasonal dynamics in 11.1.3.

River (site)	Month, year	n	NH_4^+-N $(mg\ l^{-1})$	NO_3^--N $(mg\ l^{-1})$	Total P $(mg\ l^{-1})$
Wümme	Jan. 2004	1	0.28	1.45	0.013
(peat)	Mar. 2004	1	0.44	2.66	n.d.[a]
	Apr. 2004	5	0.03	2.44	0.03
	May 2003	1	0.04	1.00	0.005
	Aug. 2004	1	0.11	0.080	0.11
	Mean	*5*	*0.17*	*1.53*	*0.04*
Ochtum	May 2003	1	0.05	1.30	0.004
(marsh)	Sept. 2004	1	0.022	0.136	0.039
	Mean	*2*	*0.036*	*0.72*	*0.02*
Elbe (gley)	May 2003	1	0.01	2.70	0.007

[a] The experiment for which the river water was taken failed because of leaking steal ring microcosms, and phosphorus was not measured. Note: this row of river water values does not belong to any of the experimental results in Table 11.2, but it was included to illustrate the seasonal dynamics of river water chemistry.

11.1.1 Site characteristics

Nutrient release (at least from peat soil, Koerselmann et al. 1993) is known to be influenced by soil parameters, vegetation, water chemistry and water level.

Original **nutrient concentrations of the river water** may influence nutrient mobilisation during floods. When comparing water samples from all three rivers (May 2003, Table 11.3), the high nitrate concentrations in the Elbe river (more than twice the concentration of the other rivers) are striking. Ammonium and phosphorus concentrations were low in all three rivers. During the inundation of a riparian meadow in Denmark, the concentration of nitrate in the river water was reduced considerably (Hoffmann et al. 2003). An important difference is the very high concentration of nitrate-N in the river Gjern studied by these authors (up to 5.7 mg l⁻¹; see Chapter 10). It is very likely that high nitrate concentrations are reduced by the flooded grassland, while low concentrations are slightly increased when the water contacts soil.

From the marsh soil, the release of nitrate-N and total P were lowest. On the same site, Erber (1998) measured a very low nitrification in the shallow ditches and explained it with the frequent inundation by stagnant water (lack of oxygen). However, she could measure nitrate at most sampling dates. This might be a result of the water flow and translocation of solved ions from the dryer ridges.

The peat released high amounts of ammonium, probably because of its high SOM content (organic matter decaying under water). Brünjes (1994) considered the import of nutrients to the "Borgfelder Wümmewiesen" by the floodwater of the river Wümme to be negligible. She found that nitrogen was only released in the form of ammonium and that mineralisation and nitrification were impeded by the high groundwater table and by the temporal inundations. Consequently, nitrate concentrations in the soil pore water were normally very low. Chemolithotrophic denitrification dominated the transformation of nitrogen. Phosphorus always occurred in very low concentrations during this study. In spite of the occasionally high mobilisation factors of P from the peat soil shown in Table 11.2, concentrations in the floodwater remained relatively low (cf. Chapters 7, 8, and 9). In a Dutch fen, P release was minor when clean Ca-rich groundwater was used for controlled flooding, while river water stimulated P release, probably due to the high sulphate content of the water (Koerselmann et al. 1993). Zak et al. (2004) found a strong retention of P by the precipitation of Fe(III) oxyhydroxides in the pore water of a peat soil. The authors only expect a P export from such peat soils when the Fe/P ratio is smaller than 3. Thus, it is very likely that the Fe/P ratio in the "Borgfelder Wümmewiesen" is lower than that, as phosphorus is released.

In a laboratory experiment with flooded peat soil cores from a Dutch fen, more N, P and K were released from *Sphagnum*-peat than from *Carex*-peat (Koerselmann et al. 1993). Although the peat soil in the present experiments was composed of *Carex gracilis* tissue and no *Sphagnum* occurred, differences in water holding capacity due to varying percentage of organic matter between soil cores may have influenced the chemistry of the floodwater. The more water was absorbed by the soil cores, the more oxygen was obviously consumed by decaying organic material of the peat and the more ammonium was released (R = 0.45** for O_2 saturation and R = -0.46* for NH_4^+-N, Spearman correlation coefficients, n = 6; cf. Chapter 9).

Compared to the other sites, ammonium release from the gley soil was very low (Table 11.2), in compliance with the low SOM content and a consequently high oxygen saturation of the flood water during the experiment. Probably nitrification is still high under these conditions and released ammonium is transformed into nitrate. The high mobilisation of nitrate, but also of phosphorus from the gley soil can probably be explained by the high input of nutrient-rich river sediments.

11.1.2 Flooding duration

In general, ammonium and phosphorus concentrations in the river water increased with flooding duration. This is shown by the dynamics experiment and corroborated when these results are compared with those of the 3-soils experiment (72h, Table 11.2). An exception is the transitory immobilisation of phosphorus in the presence of O. cyaneum in the peat soil (dynamics experiment, Table 11.2, cf. 11.2.4).

Temporal dynamics of nitrate were more complicated. For peat soil, the highest nitrate concentrations were found after one hour of flooding (in the presence of O. cyaneum, dynamics experiment, Table 11.2., Chapter 9). Also for marsh soil, the increase of nitrate concentrations was greater after 1.5 in mesocosms in the field than after 72 hours in the laboratory. Nitrification is impeded when the oxygen saturation in stagnant water is reduced. In addition, released nitrate might be reduced to ammonium when the inundation continues (re-ammonification; Bowden 1987). These results partly contradict findings of Hoffmann et al. (2003) where the reduction of nitrate concentrations in the floodwater on a riparian meadow was less pronounced when the inundation lasted for more than one day. Here again the high nitrate concentration in the river studied by these authors might cause the difference in the "biogeochemical hot moment" at the beginning of an inundation (Mc Clain et al. 2003).

In the presence of annelid worms, further influences of flooding duration arise from decomposition of faeces, gut passage, and dying animals (cf. 11.2). The short duration of flooding (maximum 72 hours as against several weeks to months in the field) was chosen to ensure the survival of the annelid worms, as the role of living individuals in nutrient release and not of their dead tissue was to be investigated. For enchytraeids, the experimental duration was already too long to survive under water. In the dead tissue experiment, the rapid decay of the tissue was observed within the 24 hours of the experiment. Thus, the "hot moment" had been matched, and a long-term experiment would not make sense in this case.

11.1.3 Season

As already mentioned, the chemical composition of river water may influence mobilisation processes from soil during flooding events. Some seasonal dynamics of river water chemistry can be derived from Table 11.3. For the river Ochtum, the reciprocal difference in nitrate and phosphorus concentrations between May and September (order of magnitude: 10) is striking. In May, there is obviously a lot of nitrification and building up of microbial biomass (i.e. incorporation of phosphorus), while in September, dying of higher organisms and leave fall start, and phosphorus is released by mineralisation. As ammonium is rapidly transformed in running waters and absorbed to the sediment, its concentration stays at a low level. The Wümme river had the highest nitrate concentrations in spring (March, April) when microbial activity (i.e. nitrification) starts. The high ammonium concentrations in January and March can be led back to organic charges of the river brought about by effluent water from the meadows.

For the marsh soil, assumptions about a season-dependent difference between the experiments in May and September should not be made because of the different flooding duration (72 vs. 1.5 hours) and different experimental conditions (laboratory – field). As the concentration of total P in the river water in January and May is very low, the calculated factor of mobilisation is high, but not the absolute amount of mobilised phosphorus.

For the peat soil, a comparison of mobilisation factors from the density-, the dead tissue- and the dynamics-experiment is possible as conditions (apart from treatments) and flooding duration (24 h) in all three experiments were the same. Although one would assume that the highest amount of ammonium and phosphorus should be released from dead tissue compared to living earthworms and their casts or animal-free controls (the main result within the dead tissue experiment), the mobilisation factor for phosphorus is much higher in the dynamics experiment, while that for ammonium is strikingly high in the density experiment. It is likely that the difference is due to the season. In January 2004 when soil cores for the dynamics experiment were taken, microbial activity and plant growth were probably absent on the site and no phosphorus was incorporated by these organisms. Also Hoffmann et al. (2004) observed a mobilisation of phosphorus in winter (field measurements). In a laboratory experiment with flooded peat soil cores, nutrient release continued at temperatures close to zero, and freezing had a stimulating effect. Nitrate release was very low in every case (Koerselmann et al. 1993). In contrast to this, Dörsch et al. (2004) did not find an effect of freeze-thaw-cycles on ammonium and nitrate pools in the field. Only N_2O emissions were increased. Although in the dynamics experiment a reactivation of microbial activity in soil cores at a laboratory temperature of 15°C was assumed, it obviously did not occur. For the density experiment, waterlogged soil cores were taken in March 2004 and additional soil cores in April. Table 11.4 shows the differences in floodwater parameters on soil cores taken in March and April. Differences between months are significant. The peak of ammonium release in spring was also described from field studies on different peat soil sites in the nature reserve "Borgfelder Wümmewiesen" (Brünjes 1994, Rodieck et al. 1992). The phenomenon was explained with the warming up of the soil, implying an increased mineralisation, while plants are not ready for the take-up of released nutrients. In contrast to this, phosphorus mobilisation from soil cores taken within the vegetation period in August 2004 (dead tissue experiment) is comparably low.

Table 11.4: Parameters of river water (Wümme) and floodwater on peat soil cores taken at different times in spring. Results of the density experiment (Chapter 8). River water: one sampling, ten measurements.

Time of sampling (in 2004)	river water March 3rd	Floodwater microcosms March 3rd	April 6th	in T/Z [a]
n	10	40	10	
oxygen satur. %	100 ± 2	35 ± 10	35 ± 8	n.s.
pH water	7.7 ± 0.1	6.2 ± 0.2	5.6 ± 0.1	T = 11.2***
electr.conduct. (μS cm-1)	471 ± 6	479 ± 40	559 ± 16	Z = -4.3***
NH_4-N (mg l^{-1})	0.03 ± 0.06	13.7 ± 3.8	4.5 ± 4.3	T = 6.7***
NO_3-N (mg l^{-1})	2.6 ± 0.5	1.2 ± 0.26	14.6 ± 6.7	Z = -4.9***
total P (mg l^{-1})	0.03 ± 0.007	0.25 ± 0.18	0.37 ± 0.18	Z = -2.7**
pH soil	-	4.9 ± 0.24	4.8 ± 0.22	Z = -2.2*

[a] Results of T-test and Mann-Whitney U-test (Z values) comparing values of microcosms with soil cores taken in March and April, respectively. * $p < 0.5$, ** $p < 0.01$, *** $p < 0.001$. Data from the dynamics experiment, Chapter 8.

11.1.4 Methodological remarks

The nutrient-protecting effect of litter described by Bowden (1987) could not occur in any of the experiments as litter was removed prior to flooding of microcosms.

In laboratory microcosms, floodwater also reaches the sides and the bottom of the soil cores. The enhanced contact surface soil-water could have had an influence on nutrient dynamics. It either supplied new surfaces (e.g., on negatively loaded clay particles) for ammonium to adsorb, reducing the concentrations in the floodwater, or the organic matter cut by the soil driller could have released more ammonium and phosphorus than undisturbed soil in the field. Not even in the field experiment natural conditions were given, as the grass turf was destroyed while inserting mesocosms into the soil. To exclude this possible influence, microcosms closing tightly around the soil cores were constructed with steel rings in the dynamics experiment, limiting the contact of the water to the surface of the core (Chapter 9, Table 11.4). No significant difference to water nutrient concentrations in PET pots (as used for all other laboratory experiments) containing the same amount of soil was found.

Thus, a more probable factor for the source of the general high release of nutrients (compared to Haslam 1998; Hoffmann et al. 2004) could be the defaunation of soil for laboratory experiments by freezing and thawing. It influences nutrient dynamics by the breakdown of microbial, plant and animal tissue (Mack 1963; Williams and Griffiths 1989; Sulkava et al. 1996). Dörsch et al. (2004) observed very high N_2O emissions from soil subjected to a freeze-thaw-cycle, leading to considerable losses of N from the system. Kampichler et al. (1999) only found a slightly higher NH_4^+-N-leaching from soil monoliths defaunated by deep-freezing and thawing compared to unfrozen controls, in some cases, also a very high release of NO_3^--N. Concentrations of ammonium and phosphorus in a field experiment on the marsh soil site (in September, i.e. without freezing) were indeed lower than in the experiment with defaunated soil in the laboratory, but this was more probably due to the shorter duration of the field experiment.

In the dynamics experiment the time for thawing of soil cores (two days) probably was too short too allow nutrient dynamics to normalise. In addition to the seasonal influence (cf. 11.1.3), this could be the cause for the high mobilisation factor of phosphorus compared to the dead tissue experiment with a similar experimental design for the controls (same duration, same soil, same water-to-soil-ratio). An extended "recovery" time for the soil cores after freezing and thawing would imply some disadvantages: The soil cores may dry out or, when they are wetted regularly, may become easily too wet. Consequently, there is formation of mould after four days (own observations). Unfortunately, no hint at the optimal recovering time of soil monoliths after thawing was found in literature.

A laboratory temperature of about 5°C during the experiments would have better met field conditions as most inundations in the study sites take place in winter. This also would have reduced the temperature contrast to the preparing freezing.

In the pots used for the laboratory experiments, the floodwater was stagnant without any exchange. In the field, an exchange, a better aeration (wind, water flow with changing water table) and thus perhaps a quicker transformation of ammonium can be assumed. Nutrients released at the beginning of an inundation could be translocated to higher zones of the wetland where there is less water movement, while there is no translocation in the stagnant water of the laboratory microcosms. However, also in the field, once a certain flood level is reached, water is standing still, often for weeks, before wind and discharge bring about movement and an increase of oxygen saturation, influencing microbial processes such as nitrification in the water. This is the case in all three study sites as dykes prevent them from being affected directly by water movement from the rivers. The peat and the marsh soil site are flooded indirectly from the surrounding ditches, while the gley soil site is a shallow depression connected to a channel (the entrance to a small harbour at the side of the river Elbe).

The presence of (growing) plants in the field experimental situation has two important effects: Released nutrients are incorporated by plants and brought to sedimentation, at least when they are adsorbed to particles. These two effects were absent in the laboratory situation.

The taking of three water samples within 24 hours in the dynamics experiment might have led to a disproportionally increased concentration of nutrients in the last water sample as the further released nutrients were added to a reduced volume of water.

11.1.5 Soil fauna

Data on nutrient mobilisation by terrestrial invertebrates are only available from terrestrial habitats, some of them at least focussing on leaching water and surface runoff after heavy rains (see Introductions of Chapters 7, 8, and 9). In aquatic habitats (lakes), different effects of sediment-dwelling macrozoobenthos on nutrient mobilisation from sediments have been observed. Fukuhara and Sakamoto (1987) recorded enhanced concentrations of ammonium in the lake water and an accelerated release of phosphorus when tubificids and/or chironomid larvae were present. In contradiction to this, Lewandowski and Hupfer (2005) measured lower concentrations of total P compared to animal-free controls. When nutrients are incorporated by microbial mats, nutrient inputs from fecal pellets of macrozoobenthos may be compensated (Tezuka 1990; Brandes and Devol 1995). However, benthivorous fish may change the whole pattern by their grazing and digging in the sediment (Breukelaar et al. 1994).

In terrestrial systems, earthworms, but also small soil-dwelling mammals like voles and moles could take the place of the fishes. Faeces are delivered by earthworms and enchytraeids as well as by other soil invertebrates. Microorganisms may play the same compensating role as in sediments. The influences of earthworms and potworms on nutrient dynamics are discussed in detail in the following paragraph.

11.2 The role of annelids in nutrient dynamics of wetlands

Numerous studies in terrestrial ecosystems have shown that earthworms and enchytraeids are involved in nutrient cycling. Their effects may be direct when nutrients are released from mucus and excrements (cf. 11.2.1) or their dead tissue (cf. 11.2.2) or when they are mobilised by bioturbation (cf. 11.2.3). It may also be indirect when the annelids foster microbial activity in soil or offer a habitat for microorganisms in their guts, initiating different processes in which these microorganisms are involved (e.g., nitrification, denitrification or immobilisation of phosphorus; cf. 11.2.4).

Fig. 11.1 shows the different steps in nitrogen cycling and phosphorus transformation where annelid worms can have an influence. These effects, as far as they have been found in the present study, are described in Table 11.5 and quantified in Table 11.6.

The following paragraphs refer to the particularities, especially of Table 11.6.

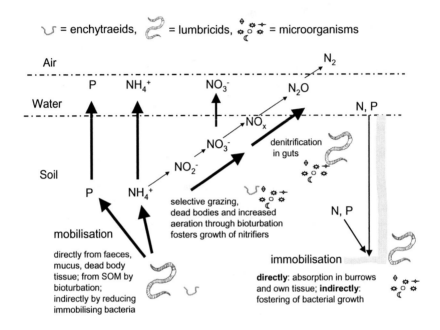

Figure 11.1: The role of earthworms and enchytraeids in nitrogen cycling and phosphorus dynamics in a terrestrial-aquatic environment. Simplified scheme developed on the base of Fig. 2.1; for sources on earthworm effects see text.

Table 11.5: Annelid effects on nutrient mobilisation in different soils. Site and animal effects summarised on the basis of the 3-soils experiment, the density experiment, the dynamics experiment, and the dead tissue experiment. Only significant effects are shown. L = earthworm (lumbricid) treatment, E = enchytraeid treatment, L+E = treatment with both groups. General mobilisation compared to river water values. Animal effects: significant differences compared to the animal-free control. For quantifying of mobilisation, see Table 11.2, for quantification of annelid effects see Table 11.6.

Soil	Mobilisation of...	Animals mobilise...	No effect ...
peat	ammonium phosphorus	L: phosphorus, ammonium L+E: phosphorus **dead L tissue**: phosphorus, ammonium	of annelids on nitrate, but: less nitrate in the presence of dead L tissue
marsh soil	phosphorus	L+E: phosphorus, nitrate	of L or E alone
gley soil	phosphorus ammonium nitrate nitrite	L+E: nitrite	of L or E alone, but phosphorus tends to be lower in E treatment (n.s.)

Table 11.6: Earthworm effects on nutrient release from peat soil calculated for the field scale. % compared to an animal-free control (= 100%). No absolute values (in kg ha^{-1}) are given because of the methodological shortcomings (cf. 11.1.4). From the relative values, the following trends can be derived: The earthworm effect on nutrient mobilisation differs in different soils (e.g. high mobilisation factor of nitrate-N only in the gley soil). The earthworm effect on mobilisation of ammonium and total phosphorus is more pronounced after a longer time of flooding and at a higher density. The effect of *O. cyaneum* is more pronounced than that of *L. rubellus*, and the amount of nutrients mobilised from dead tissue is higher than that mobilised by living earthworms in the same time.

Experiment				Earthworm species	Mio. ind. ha^{-1}	Biomas s (kg ha^{-1})	Earthworm effect (% compared to control)		
Ch.	Name/ soil	h	n				NH$_4^+$-N	NO$_3^-$-N	Total P
7	3 soils/ marsh	72	5	*L. rubellus*	6	972	212	80	197
	3 soils/ gley^b	72	5	*L. rubellus*	6	1270	21	122	107
	3 soils/ peat	72	5	*O. cyaneum*	6	2452	286	103	236
8	density/ peat	24	4	*O. cyaneum*	2	1473	122	129	200
9	dynamics/ peat	24	5	*O. cyaneum*	8	1326	192	217	248
	dead tissue/ peat	24	6	*L. rubellus*	2	1294	113	77	129
		24	6	*L. rub.* + *L. rub.* dead tissue	2 + 4	3144	229	32	212
		24	6	*L. rub* + *O. cyan.* dead tissue	2 + 4	3025	218	35	243

^{b} additionally, only for the gley soil: 257% of NO$_2^-$-N (nitrite) in the presence of *L. rubellus*.

11.2.1 Excrements

It is fair to assume that most of the enhanced ammonium and phosphorus concentrations in the presence of living earthworms (and enchytraeids) originate from the animals' excrements (Haimi 1995). Fresh earthworm casts as well as enchytraeid excrements are known to be rich in NH_4^+-N, NO_3^--N, and phosphorus (earthworms: Parle 1963, Graff 1970, Sharpley and Syers 1976, 1977, Scheu 1987a, Haimi and Huhta 1990, Elliott et al. 1991, Parkin and Berry 1994, Aira et al. 2003, Haynes et al. 2003; enchytraeids: Williams and Griffith 1989, Schrader et al. 1997). Table 11.7 shows some literature values for nutrient contents of earthworm casts.

Table 11.7: Nutrient contents of fresh annelid casts. Increase = factor of increase compared to uningested soil. n.d. = no data; n.s. = no significant difference.

Species	Nutrient	Content ($\mu g\ g^{-1}$ dw)	Increase	Source
O. tyrtaeum	NO_3^-	5 – 15	n.s.	Parkin & Berry 1994
	NH_4^+	40 –70	1.3 – 2.3	
L. rubellus,	NH_4^+-N	54-110	2.8 – 5.5	Syers et al. 1979
A. caliginosa	NO_3^--N	15-48	1 – 3.2	
	Total N [a]	38	1.7	
A. caliginosa	Total N	n.d.	n.s.	Scheu 1987a
A. caliginosa	Olsen P	24	1.2	Haynes et al. 2003
	NH_4^+-N	80	4	
	NO_3^-N	125	5	
L. rubellus	Olsen P	30	1.5	
	NH_4^+-N	180	9	
	NO_3^-N	150	6	
L. rubellus,	NH_4^+-N	1.7 – 3.2	0 – 1.3	Elliott et al. 1991
A. caliginosa	NO_3^-N	3 - 11	0 – 5	
A. caliginosa,	Total P	83	n.s.	Le Bayon & Binet
L. terrestris				1999
Enchytraeus	N_t	2.4	1.6	Schrader et al. 1997
spec.				

[a] means of all seasons

Earthworms normally deposit their casts on the soil surface (Shipitalo et al. 1988) where they are directly exposed to flood water. In a study by Le Bayon and Binet (2001), the erodibility of newly deposited casts by rainfall was high. Amounts of sediment and nutrient losses in the runoff were at least twice as high without as in the presence of surface casts due to a surface roughness produced by earthworm activity (burrows and casts). Soil particles and phosphorus were transferred over a short distance by suspension and deposition.

In all experiments carried out here, the earthworms had emptied their guts on filter paper in the night preceding the inoculation of microcosms. The uptake of food and the gut passage take 6 to 8 h (Daniel and Anderson 1992). Thus, after 24 hours inoculation time, the presence of fresh casts was expected and observed on the surface of soil cores in all experiments. As newly formed casts were quickly disintegrated, counting of casts like in terrestrial experiments was not possible. After the 72 hours (3 soils experiment), a greater number of casts was probably present, and concentrations of ammonium and phosphorus in peat soil microcosms were indeed higher than after 24 hours (see Table 11.2: the mobilisation factor for NH_4^+ was 24-164 after 24 h versus 387 after 72 h; for total phosphorus, the mobilisation factor was 1.2 – 48 after 24 h versus 118 [143 with L and E] after 72h).

The proportional additional release of ammonium and total phosphorus in the presence of earthworms correlated strongly positive (Pearson correlation coefficient $R = 0.877$, $p < 0.01$, $n = 8$; calculated with mean values from all replicates of each experiment, cf. Table 11.6). This hints at a common source of these two nutrients, very likely the earthworms' faeces. The more nutrients (and other substances that were not measured) are released into the water from the earthworms' excretes, the higher the electrical conductivity and the lower the oxygen saturation of the water. This could also cause death and decay of enchytraeids and thus a further release of NH_4^+ and P from their dead tissue (cf. 11.2.2).

The initially high amount of ammonium in earthworm casts decreases due to nitrification (Parle 1963; Scheu 1987a; McInerney and Bolger 2000; Haynes et al. 2003), usually down to trace amounts within some days (Syers et al. 1979; Martin and Marinissen 1993). Under water, nitrification is impeded (Bowden 1987). Thus, ammonium released from casts is not nitrified and therefore, it accumulates.

Especially endogeic earthworms ingest large amounts of soil (Scheu 1987b) and fragmented litter material (Scheu and Sprengel 1989). Ruz-Jerez et al. (1992) found that by the earthworms' consumption of litter, a greater proportion of carbon is oxidised than in the presence of microbial decomposers alone. Hence, a greater proportion of the nutrients stabilised in organic matter (N, P S) is liberated. *L. rubellus* produces high amounts of casts in relation to its body weight (80 to 460 mg g^{-1} live worm day^{-1} vs. 70 to 180 mg g^{-1} live worm day^{-1} for *L. terrestris*, Shipitalo et al. 1988). Temperature may have an influence on the amount of produced casts, at least for *L. terrestris*. Hartenstein and Amico (1983) collected 250 mg g^{-1} live worm day^{-1} in an experiment carried out at 23°C (compared to 15°C in the present as well as in the experiment carried out by Shipitalo et al.). However, there are differences between species, and it is not sure if temperature has the same influence on casting activity of the earthworms used in the present study. In addition, season may determine the amount of casts, as earthworms are more active during wet periods (Sharpley and Syers 1977).

The mean annual production of surface casts amounted to 25-33 tonnes ha^{-1} y^{-1} for *L. rubellus* in a New Zealand pasture (Sharpley and Syers 1977) and to 34 kg (dry weight) year^{-1} kg^{-1} earthworm fresh weight in a corn field with *A. caliginosa* and *L. terrestris* (Le Bayon and Binet 1999). In a field experiment in a cornfield, rates of cast production were higher when the matrix potential of the soil was higher, probably because of a greater activity of the earthworms. Consequently, the amount of water per day egested by the earthworms (*A. rosea, A. caliginosa,* and *A. trapezoides*) increased. However, the amount of water egested per cast remained constant (Hindell et al. 1994). It remains unclear if under flooding conditions, earthworm activity is fostered or if the animals are impeded, at least when the oxygen saturation is lowered.

Phosphorus losses from cast erosion in a cornfield during a rainfall event amounted to 25-49 mg phosphorus m^{-2}, depending on soil compaction (Le Bayon and Binet 1999). It is very likely that phosphorus contents of earthworm casts and phosphorus losses are rather lower in non-fertilised grassland like in the present study than in this cornfield. Phosphatase activity in earthworm casts is high and organically bound P is transformed into inorganic P. Hence, it can be easily washed out (Sharpley and Syers 1976, Satchell and Martin 1984, Scheu 1987a, Krishnamoorthy 1990).

In the beginning, nutrients in annelid casts are occluded by the compact structure of the casts and only leached later on (Wolters 1988). McInerney and Bolger (2000) at first did not find an effect of casts on ammonium concentrations of leachates. However, when soil and casts were water-saturated, the nutrient-protecting effect of earthworm casts obviously was reduced and mineral N was leached. Binet and Le Bayon (1999) reported a much shorter lifetime of fresh casts under wet conditions (four days) than during a dryer period (up to 14 days). However, in their study casts did not disappear at the first rain event but only during the second and third rain event. In the present experiment, this could be another explanation for the retarded release of ammonium in spite of the presence of casts at the beginning of flooding.

In another study of Le Bayon and Binet (1999), newly formed earthworm casts disappeared quicker than older ones during rainfall events. Disruption of casts was due to the splash effect and not to the runoff effect. Under the experimental conditions of flood simulation, there was no splash effect, and only fresh earthworm casts were present.

Earthworm casts absorb H^+ ions and their mucus can neutralise acidic as well as alkaline environments (Schrader 1994). Consequently, the water pH of the treatments with *O. cyaneum* was somewhat less acidic than in the control after 24 h (dynamics experiment, Chapter 9). This could indirectly influence pH-dependent nitrogen transformation processes (e.g., the balance between NH_4^+ and NH_3).

Enchytraeids use earthworm casts as a food resource, promoting the decomposition of enclosed litter (Zachariae 1967). This could lead to the enhanced mobilisation in the presence of both groups.

11.2.2 Dead tissue

A mobilisation of ammonium and phosphorus from dead earthworm tissue has been found repeatedly (Christensen 1987; Curry 1994; Weiß 1994). For the flooding situation, this was partly proved in the first experiment (positive correlation of the number of dead *L. rubellus* and phosphorus concentrations in the marsh and gley soil microcosms) and confirmed in the dead tissue experiment (considerably higher phosphorus and ammonium concentrations in the presence of dead earthworm tissue). The amount of nutrients released from the tissue of dead enchytraeids is probably too small to be of any importance (Williams and Griffiths 1989), but could have influenced the significant increase of NH_4^+ and total phosphorus concentrations in the presence of earthworms in the 3-soils experiment (peat and marsh soil) as well as nitrite concentrations (gley soil). Unfortunately, nitrite was not measured in the dead tissue experiment.

The tissue of earthworms is quickly mineralised after death (Curry 1994). At a temperature of 12°C, earthworm tissue was completely mineralised within three weeks. After one month, 75% of the N previously stored in the earthworm tissue was found back in soil (Christensen 1987). However, it is possible that the freezing and thawing of the earthworms in the present experiment caused an even faster mineralisation of their tissue than after the natural death (Mack 1963, cf. 11.1.4).

The dead tissue experiment shows that also the release of salts (measured in the higher electrical conductivity) and the oxygen consumption by microorganisms are processes related to the decay of dead (earthworm) tissue.

Nitrate concentrations in the floodwater were reduced considerably in the presence of dead earthworm tissue (Table 11.6). Under micro-aerobic conditions, nitrification is impeded, and two nitrate-reducing processes carried out by (facultatively) anaerobic bacteria may take place: denitrification (Well et al. 2002; Hefting 2003) and dissimilatory nitrate reduction (Atlas and Bartha 1998; cf. Chapter 2).

11.2.3 Bioturbation

It is possible that the higher burrowing activity of the endogeic *O. cyaneum* compared to the epigeic *L. rubellus* was the cause of its greater effect on ammonium and phosphorus mobilisation (Chapter 9, Table 11.6: comparison of the density experiment with the dead tissue experiment; same duration and same density of earthworms). The increased N_{min} leaching (up to a factor of 2.0) in the presence of the earthworm *O. lacteum* found by Scheu (1993, 1995) was also explained with the burrowing activity of the endogeic earthworm.

The burrows of the anecic *L. terrestris* have a positive effect on leaching losses from agroecosystems as they increase the leachate volume (Nagel and Beese 1992), resulting in a 2.5-fold loss of inorganic N (Subler et al. 1997; Domínguez et al. 2004). Especially concentrations of DON in leachates from sites with *L. terrestris* were increased (Domínguez et al. 2004). In the present study, DON was not measured, only the inorganic ammonium and nitrate. Thus, the N-mobilising earthworm effect in a terrestrial-aquatic system could still be greater than measured until now. However, it is unclear if the epigeic *L. rubellus* or the endogeic *O. cyaneum* used for the present experiments have the same effect as *L. terrestris*.

11.2.4 Microbial activity

In many experiments and field studies on nutrient dynamics in terrestrial ecosystems, an indirect effect of annelid worms by fostering microorganisms involved in mineralisation and nitrification has been observed.

In this study, NO_3^- concentrations in the floodwater were increased in the presence of annelids and correlated negatively with the number of dead enchytraeids in the marsh and the gley soil (3 soils experiment, Chapter 7). This hints at a role of living enchytraeids in nitrification. Like other mesofauna organisms, enchytraeids play a regulating role in mineralisation processes through their selective grazing on microorganisms (Dash 1972; Didden 1993). They can increase the leaching of dissolved organic nitrogen, ammonium, and phosphorus (Williams and Griffiths 1989; Briones et al. 1998).

The positive correlation of ammonium and water pH in the presence of enchytraeids and *L. rubellus* in the peat and the marsh soil could be explained by the sorption of H^+-ions on microorganisms or SOM provided by the (dead) animals.

The accumulation of nitrite in the gley soil could also have been caused by the enchytraeids fostering the microorganisms involved in the first step of nitrification. Obviously, there is a lot of nitrification in the presence of earthworms in the beginning of an inundation (dynamics experiment with peat soil, Chapter 9), perhaps by breaking up litter and/or fostering microorganisms. Later on, the effect weakens, although the mobilisation of nitrate is still higher than in the control treatment.

Microorganisms like those in the earthworms' guts, in their faeces or on their dead tissue can be source and sink for nitrogen (Paul and Clark 1989). Nagel (1996) found an immobilisation of soil nitrogen and an increased release of N from straw litter to leachates in the presence of earthworms. In another experiment, *L. terrestris* initially reduced respiration as well as the release of straw-borne N by increased immobilisation. The authors assumed that the earthworms cause a longer lasting, more continuous mineralisation by the microflora ("buffering effect"; Nagel et al. 1995).

In the field, both fractions of N_{min}, ammonium and nitrate, are released in higher concentrations in the presence of earthworms. Living earthworms, but also enchytraeids foster microorganisms involved in nitrification (Scheu 1993; Weiß 1994; Didden 1995; Haimi 1995; Haimi and Huhta 1995; Helling 1997). Earthworms fostered N-mineralisation in a field experiment with manipulated densities in a cornfield. However, it was not clear if the increased N mineralisation was due to an increased mobilisation of N or due to decreased losses of N from the surface soil (Subler et al. 1998). Blair et al. (1997) measured a significantly lower microbial biomass N in treatments with increased earthworm abundance, while extractable NH_4^+ and NO_3^- concentrations were higher, at least in inorganically fertilized systems. The authors assume that earthworms may increase N availability by reducing microbial immobilisation and enhancing mineralisation. Marinissen and

De Ruiter (1993) estimated the total (direct and indirect) contribution of earthworms to N mineralisation in a no-tillage agroecosystem to amount to 363 kg ha^{-1} year^{-1}.

Next to the direct relationship between earthworms and microorganisms via the earthworms' guts as a habitat or their excretes (faeces, mucus) as a substrate to grow in, there are three other phenomena related to earthworm activity that influence microorganism growth: (i) soil bioturbation enhancing soil aeration and the utilization of soil organic matter by microorganisms (cf. 11.2.3); (ii) stimulating effects of worms on protozoa which in turn may increase bacterial activity; (iii) chemical mediators that act at low concentrations on microbial metabolism (Binet et al. 1998). The authors found a 3- to 19-fold increase of protozoan population density in the presence of *L. terrestris*. However, there was no correlation between the increase in microbial respiration and earthworm weight. The authors suggest that microbial utilisation of earthworm excreta is favoured by catalytic substances like vitamins in mucus or casts. This corresponds to the observation made by Haynes et al. (2003) that casts of *L. rubellus* and *A. caliginosa* contained a smaller, but more metabolically active microbial community than the surrounding soil.

In the density experiment, the negative correlation of oxygen saturation with total P in the presence of *O. cyaneum* could be less pronounced than in treatments with enchytraeids because the earthworms foster oxygen-consuming and P-limited microorganisms (Karsten 1997) by providing phosphorus in their casts. The same is true for microorganisms living in the earthworm guts. The reduced phosphorus concentration in the presence of earthworms after 5 hours followed by an increased concentration after 24 hours hint at a temporary immobilisation of P by the gut flora. Later, when casts are mineralised and organic phosphorus turns into the inorganic form (Satchell and Martin 1984; Scheu 1987a), concentrations in the water rise again.

L. terrestris reduces microbial biomass and nitrifying activity, while it stimulates proteolytic bacteria (Devliegher and Verstraete 1995). The authors differentiate between nutrient-enrichment processes (associated with the organic matter incorporation, resulting in an increased nutrient concentration in casts and surrounding soil, thus increasing microbial biomass and activity) and gut-associated processes i.e. the transit of soil and organic matter through the earthworm gut (reducing microbial biomass by digestion or by reducing the substrate).

Simek and Pizl (1991) measured a higher nitrogenase activity in guts and casts as well as on the skin of *L. rubellus* and *A. caliginosa*, although populations of N_2-fixing microorganisms were less dense. The authors assume that microbial activity was enhanced by the rich source of carbon compared to the surrounding soil. Striganova et al. (1993) measured a 3-fold nitrogenase activity in the guts of *L. terrestris* compared to its food (litter), while they did not find any difference for *L. rubellus* and *A. caliginosa*.

11.2.5 Annelid density and activity

A simulated earthworm density of 600 ind. m^{-2} showed clear effects on nutrient mobilisation from all three soils (Chapter 7, Table 11.2, and Table 11.6). The highest simulated density, 800 ind. m^{-2} of *O. cyaneum*, had an even stronger effect on phosphorus, but not on ammonium; for the latter, the longer duration of the 3-soils-experiment obviously was more decisive. Probably the decrease of oxygen saturation in the water favoured the accumulation of ammonium, while the release of phosphorus is independent of aeration. Also a density of 200 ind. m^{-2} of *O. cyaneum* still had a significant effect on the mobilisation of both nutrients (Table 11.6), while *L. rubellus*, inoculated at the same density in the same soil, had no effect, probably due to its weaker bioturbating activity (Chapter 9, 11.2.3). For the peat soil experiments, a positive correlation between phosphorus concentrations and earthworm biomass (including the fresh weight of dead earthworms in the dead tissue experiment) was found (Pearson correlation coefficient R = 0.8, p < 0.05, n = 6, calculated with the percentage of increase compared to the animal-free controls shown in Table 11.6).

A density of 600 ind. m^{-2} is within a maximum range for the gley soil, whereas for the other soils, such a density was never observed during the field investigations (maximum abundance: gley soil 696 ± 66 ind. m^{-2}, marsh soil 156 ± 91 ind. m^{-2}, peat soil 67 ± 52 ind. m^{-2}). The highest density recorded during field investigations on the peat soil site was 100 ind. m^{-2} for *L. rubellus* or *O. cyaneum* alone and 140 ind. m^{-2} for all earthworms (Chapter 5). These are much lower densities than used in the laboratory experiments. Thus, the mobilising effect of earthworms on the study site itself is probably negligible, though in other sites, much higher densities were found, and an effect of earthworms has to be taken into account (Graefe 1998; Zorn 2004a, b).

Sulkava et al. (1996) found a positive correlation between the amount of NH_4^+ in soil and enchytraeid biomass. The simulated enchytraeid density of 6000 ind. m^{-2} was within the lower range of mean field abundances (gley soil 4900 ± 5500 ind. m^{-2}, marsh soil $13,000 \pm 11,000$ ind. m^{-2}, peat soil $8100 \pm 11,000$ ind. m^{-2}). However, at favourable soil moisture conditions, a much higher maximum abundance of enchytraeids (20,000 to 50,000 ind. m^{-2}) occurred in all three soils (Chapter 5). A possible effect of enchytraeids at these high densities can be assumed regarding the effects observed at the simulated lower density. In general, density-dependent effects of enchytraeids were hardly found in the respective experiment, and if so, they varied over time. Next to the low density, the very short incubation time, followed by the high mortality or inactivity of enchytraeids when flooded, is probably the main reasons for this.

11.3 Conclusions and recommendations for water management

Wetland management often aims at providing habitat to endangered wading birds (Ausden et al. 2001). Earthworms (and probably also the bigger enchytraeids, A. Schoppenhorst, pers. comm.) are an important food source for these birds. Therefore, controlled flooding of wetlands should aim at stabilising their populations. Another purpose of wetlands is the reduction of nutrient concentrations in the river water.

Extended flooding from late autumn until spring reduces annelid populations, i.e. the food resource for wading birds. Next to this, considerable amounts of ammonium and phosphorus are released from dead earthworm tissue. Inundations during the hibernal diapause of annelid worms will be less problematic. Also during short-term inundations of less than one day, as they occur in tide-influenced floodplains, the mobilising effect of earthworms is negligible and survival is high. The soil should be allowed to drain in spring. During the breeding season when earthworms are needed as food resource, the water table can be raised artificially again to make earthworms concentrate in the upper soil layer were they are available for birds (without flooding the meadows as this would impair breeding birds). Also in summer, a higher water table will minimise negative effects of summer drought on earthworm populations (Ekschmitt 1991), thus ensuring the food resource for the next bird generation, and minimising undesired nutrient release into the river.

11.4 Research needs

The following research needs remain:

- The role of one species in different soils and of different species in the same soil should be compared systematically.

- The development of nutrient release under the influence of annelids in the course of an inundation lasting several weeks should be observed in a long-term experiment.

- The role of enchytraeids in peak densities should be investigated with the help of suited microcosms. An experiment with smaller soil volumes would facilitate the simulation of high densities. Living individuals as well as the tissue of dead enchytraeids should be used for the inoculation of different microcosms.

- The laboratory results should be validated under field conditions, if possible in a meadow where controlled flooding or at least a raise of the groundwater table is feasible for the experiment.

- The possible processes involved in the reduced nitrate concentrations in the presence of dead earthworm tissue (enhanced denitrification, reduced nitrification, and re-ammonification) should be resolved experimentally.

12. References

Aira M, Monroy F & Domínguez J (2003) Effects of two species of earthworms (*Allolobophora* spp.) on soil systems: a microfaunal and biochemical analysis. Pedobiologia 47: 877-881

Atlas RM & Bartha R (1998) Biogeochemical cycling: nitrogen, sulfur, phosphorus, iron and other elements. In: Atlas, R. M., Bartha, R. (Eds.), 4, Microbial ecology, fundamentals and applications Benjamin Cummings Science Publishers, Menlo Park, California

Augustin J, Merbach W & Russow R (1997) Einfluss von Rohrglanzgras (*Phalaris arundinacea* L.) auf N-Umsetzungsprozesse und die Emission klimarelevanter Spurengase in Modellversuchen mit Niedermoorsubstrat. In: Merbach, W. (Ed.), Rhizosphärenprozesse, Umweltstress und Ökosystemstabilität. 7. Borkheider Seminar zur Ökophysiologie des Wurzelraumes (pp 101-108). B.G. Teubner Verlagsgesellschaft, Stuttgart, Leipzig

Ausden M, Sutherland WJ & James R (2001) The effects of flooding lowland wet grassland on soil macroinvertebrate prey of breeding wading birds. Journal of Applied Ecology 38: 320-338

Beylich A & Graefe U (2002) Annelid coenoses of wetlands representing different decomposer communities. In: Broll, G., Merbach, W., Pfeiffer, E.-M. (Eds.), Wetlands in Central Europe (pp 1-10). Springer, Berlin

Binet F, Fayolle L, Pussard M, Crawford JJ, Traina SJ & Tuovinen OH (1998) Significance of earthworms in stimulating soil microbial activity. Biology and Fertility of Soils 27: 79-84

Binet F & Trehen P (1992) Experimental microcosm study of the role of Lumbricus terrestris (Oligochaeta: Lumbricidae) on nitrogen dynamics in cultivated soils. Soil Biology and Biochemistry 24: 1501-1506

Binet F & Le Bayon R-C (1999) Space-time dynamics in situ of earthworm casts under temperate cultivated soils. Soil Biology and Biochemistry 31: 85-93

Blair J, Parmelee RW, Allen MF & McCartney DA (1997) Changes in soil N pools in response to earthworm population manipulations in agroecosystems with different N sources. Soil Biology and Biochemistry 29: 361-367

Blanchart E, Frenot Y & Tréhen P (1987) Signification biologique du potentiel hydrique dans la distribution des Diptères à larves hydrophiles. Pedobiologia 30: 333-344

Blanken-Mittendorf W (1990) Elektrischer Regenwurmfang mit der Oktett-Methode nach Thielemann - Methodenüberprüfung und Erfassung der

Lumbricidenfauna auf landwirtschaftlich genutzten Hochmoorböden. Diploma thesis, University of Bremen, Bremen

Bobbink R & Lamers LPM (2002) Effects of increased nitrogen deposition. In: Bell, J. N. B., Treshow, M. (Eds.), Air pollution and plant life

Bohlen PJ, Parmelee RW, Blair JM, Edwards CA & Stinner BR (1995) Efficacy of methods for manipulating earthworm populations in large-scale field experiments. Soil Biology and Biochemistry 27: 993-999

Bolte D & Moritz M (1988) Ermittlung des Ist-Zustandes (Besiedlungsdichten und Artenspektren) ausgewählter Bodentiergruppen im Gebiet von Brokhuchting, Bremen. Report for Senator für das Bauwesen, Gartenbauamt. Bremen

Borcherding F (1889) Das Tierleben auf und an der "Plate" bei Vegesack. Abhandlungen des Naturwissenschaftlichen Vereins zu Bremen 11: 265-279

Borek V, Morra MJ, Brown PB & McCaffrey JP (1995) Transformation of the glucosinolate-derived allelochemicals allyl isothiocyanate and allylnitrile in soil. Journal of Agriculture and Food Chemistry 43: 1935-1940

Bouché MB (1977) Stratégies lombriciennes. Ecological Bulletin 25: 122-132

Bowden WB (1987) The biogeochemistry of nitrogen in freshwater wetlands. Biogeochemistry 4: 313-348

Brünjes H (1994) Ökochemische Untersuchungen zur Nährstoffsituation an ausgewählten Standorten der Borgfelder Wümmewiesen. Diploma thesis, University of Bremen, Bremen

Brandes JA & Devol AH (1995) Simultaneous nitrogen and oxygen respiration in coastal sediments: Evidence for discrete diagenesis. Journal of Marine Research 53: 771-797

Brandsma OH (2002) Die Bedeutung der Düngung für das Nahrungsangebot von Wiesenvögeln. In: Düttmann, H., Ehrnsberger, R., Faida, I. (Eds.), Wiesenvogelschutz in Norddeutschland und den Niederlanden, Symposium an der Hochschule Vechta, Abstracts (pp 67-68). Vechta

Brandsma OH (1992) Onderzoek weidevogelbeheer en bodemfauna in her reservaatsgebied Giethoorn-Wanneperveen V. Directie beheer landbouwgronden - landbouw, natuurbeheer en visserij, Zwolle

Brandsma OH (1997) Onderzoek weidevogelbeheer en bodemfauna in het reservaatsgebied Giethoorn-Wanneperveen VIII (1992-1996). Directie beheer landbouwgronden - landbouw, natuurbeheer en visserij, Zwolle

Brauns A (1954) Terricole Dipterenlarven - Eine Einführung in die Kenntnis und Ökologie der häufigsten bodenlebenden Zweiflüglerlarven der Waldbiozönose auf systematischer Grundlage. Musterschmidt-Verlag, Göttingen

Breukelaar AW, Lammens EHRR, Breteler JGPK & Tátrai I (1994) Effects of benthivorous bream (*Abramis brama*) and carp (*Cyprinus carpio*) on sediment resuspension and concentrations of nutrients and chlorophyll a. Freshwater Biology 32: 113-121

Briones MJI, Carreira J & Ineson P (1998) *Cognettia sphagnetorum* (Enchytraeidae) and nutrient cycling in organic soils: a microcosm experiment. Applied Soil Ecology 9: 289-294

Briones MJI, Ineson P & Piearce TG (1997) Effects of climate change on soil fauna: responses of enchytraeids, Diptera larvae and tardigrades in a transplant experiment. Applied Soil Ecology 6: 117-134

Brodmann PA & Reyer H-U (1999) Nestling provisioning in water pipits (*Anthus spinoletta*): Do parents go for specific nutrients or profitable prey? Oecologia 120: 506-514

Cejka T 2003, The molluscan fauna along the moisture gradient on the lower Morava river, Slovakia: Sbornik prirodovedneho klubu v Uherskem Hradisti (in press)

Cejka T (1999) The terrestrial molluscan fauna of the Danubian floodplain (Slovakia). Biologia 54: 489-500

Chan KY & Munro K (2001) Evaluating mustard extracts for earthworm sampling. Pedobiologia 45: 272-278

Christensen O (1987) The effect of earthworms on nitrogen cycling in arable soils. Proceedings of the 9th International Colloquium on Soil Zoology 1985: 106-118

Cowardin LM, Carter V & LaRoe ET (1979) Classification of wetlands and deepwater habitats of the United States. U.S. Department of the Interior, Washington D.C.

Curry JP (1994) Grassland invertebrates - ecology, influence on soil fertility and effects on plant growth. Chapman&Hall, London

Curry JP & Boyle KE (1987) Growth rates, establishment and effects on herbage yield of introduced earthworms in grassland on reclaimed cutover peat. Biology and Fertility of Soils 3: 95-98

Düttmann H & Emmerling R (2001) Grünland-Versauerung als besonderes Problem des Wiesenvogelschutzes auf entwässerten Moorböden. Natur und Landschaft 76: 262-269

Dahl A, Klemm M & Weis M (1993) Untersuchung der Land-Gehäuseschnecken im Rahmen des Pflege- und Entwicklungsplanes für die Fischerhuder Wümmeniederung. Report for Biologische Station im Kreis Osterholz (BiOS). Tübingen

Daniel O & Anderson JM (1992) Microbial biomass and activity in contrasting soil materials after passage through the gut of the earthworm *Lumbricus rubellus* Hoffmeister. Soil Biology and Biochemistry 24: 465-470

Dash MC & Cragg JB (1972) Selection of microfunghi by Enchytraeidae (Oligochaeta) and other members of the soil fauna. Pedobiologia 12: 282-286

Davison L, Headley T & Pratt K (2004) Aspects of design, structure, performance and operation of reed beds - eight years experience in north eastern New South Wales. Water Science and Technology

Delettre YR (2000) Larvae of terrestrial Chironomidae (Diptera) colonize the vegetation layer during the rainy season. Pedobiologia 44: 622-626

Deutscher Wetterdienst (DWD) (2003) Das Niederschlagsgeschehen in Mitteleuropa in den ersten 12 Tagen des August 2002 im Vergleich zum klimatologischen Mittel 1961-1990. Klimastatusberichte des DWD, Offenbach a. M.

Deutscher Wetterdienst (DWD) (2004) Der Rekordsommer 2003. Klimastatusberichte des DWD, Offenbach a. M.

Devliegher W & Verstraete W (1995) *Lumbricus terrestris* in a soil core experiment: nutrient enrichment processes (NEP) and gut-associated processes (GAP) and their effect on microbial biomass and microbial activity. Soil Biology and Biochemistry 27: 1573-1580

Didden WAM (1993) Ecology of terrestrial Enchytraeidae. Pedobiologia 37: 2-29

Didden WAM (1995) The effect of nitrogen deposition on enchytraeid-mediated decomposition and mobilization - a laboratory experiment. Acta Zoologica Fennica 196: 60-64

Didden WAM, Born H, Domm H, Graefe U, Heck M, Kühle J, Mellin A & Römbke J (1995) The relative efficiency of wet funnel techniques for the extraction of Enchytraeidae. Pedobiologia 39: 52-57

Dörsch P, Palojärvi A & Mommertz S (2004) Overwinter greenhouse gas fluxes in two contrasting agricultural habitats. Nutrient Cycling in Agroecosystems 70: 117-133

Dohle W, Bornkamm R & Weigmann G (1999) Das Untere Odertal: Auswirkungen der periodischen Überschwemmungen auf Biozönosen und Arten. E. Schweizerbart'sche Verlagsbuchhandlung, Stuttgart

Dunger W & Fiedler HJ (1989) Methoden der Bodenbiologie. Gustav Fischer Verlag, Stuttgart

East D & Knight D (1998) Sampling soil earthworm populations using household detergent and mustard. Journal of Biological Education 32: 201-206

Edwards CA & Lofty JR (1977) Biology of earthworms. Chapman&Hall, London

Edwards WM (1992) Role of *Lumbricus terrestris* (L.) burrows on quality of infiltrating water. Soil Biology and Biochemistry 24: 1555-1561

Ehrmann O & Babel U (1991) Quantitative Regenwurmerfassung – ein Methodenvergleich. Mitteilungen der Deutschen Bodenkundlichen Gesellschaft 66: 475-478

Eichinger E (2004) Nebeneffekte der Regenwurmextraktion aus dem Boden mittels Formalin-Methode. Diploma thesis, Universität für Bodenkultur (BOKU), Wien

Eidgenössische Forschungsanstalten FAL FR (2002) Bestimmung der Regenwurmpopulation (Biomasse, Abundanz). In: Anonymous, Band 2: Bodenuntersuchung zur Standort-Charakterisierung (pp 1-3)

Ekschmitt K (1991) Abschlussbericht über Begleituntersuchungen zur Wirkung unterschiedlicher Überstauungsintensitäten auf die Bodenfauna von Grünland 1987-1990. Report for Senator für Umwelt und Stadtentwicklung, Bremen

Elliott PW, Knight D & Anderson JM (1991) Variables controlling denitrification in earthworm casts. Biology and Fertility of Soils 11: 24-29

Emmerling C (1995) Long-term effects of inundation dynamics and agricultural land-use on the distribution of soil macrofauna in fluvisols. Biology and Fertility of Soils 20: 130-136

Emmerling C (1993) Methodenvergleich zur Eignung von Senf als Extraktionsmittel. Mitteilungen der Deutschen Bodenkundlichen Gesellschaft 75: 133-136

Emmerling C (1993) Nährstoffhaushalt und mikrobiologische Eigenschaften von Auenböden sowie die Besiedlung durch Bodentiere unter differenzierter Nutzung und Überschwemmungsdynamik. PhD thesis. Verlag Shaker, Aachen, Trier

Engelhardt W (1986) Was lebt in Tümpel, Bach und Weiher? Pflanzen und Tiere unserer Gewässer, kosmos Naturführer. Kosmos, Stuttgart

Erber C (1998) Bodeneigenschaften und Stoffhaushalt winterlich überstauter Flussmarschen des Niedervielandes bei Bremen. PhD thesis Justus-Liebig-Universität, Gießen

Erber C, Felix-Henningsen P, Handke K, Kundel W & Schreiber K-F (2002) Management of moist grassland in a fresh-water marsh of the Weser river: effects on soil, vegetation and fauna. In: Broll, G., Merbach, W., Pfeiffer, E.-M. (Eds.), Wetlands in Central Europe (pp 71-98). Springer, Berlin

Evans AC & Guild WJ (1948) Studies on the relationships between earthworms and soil fertility. Ann Appl Biol 79: 95-108

Faber JH, van Kats RJM, Aukema JM, Bodt JM, Burgers J, Lammertsma DR & Noordam AP (1999) Ongewervelde fauna van ontkleide uiterwaarden. Instituut voor Bos- en Natuuronderzock, Wageningen

Faber JH, Burgers J, Aukema JM, Bodt JM, van Kats RJM, Lammertsma DR & Noordam AP (2000) Ongewervelde fauna van ontkleide uiterwaarden - Monitoringverslag 1999. Alterra, Research Instituut voor de Groene Ruimte, Wageningen

Faida I, Düttmann H & Ehrnsberger R (2003) Evaluation zum Symposium Wiesenvogelschutz in Norddeutschland und den Niederlanden in Vechta 2002. Verlag Druckerei Runge, Cloppenburg

Filser J, Mommertz S, Ackermann G & Ehrengruber A, 1996, Großpraktikum Ökologie: Regenwurmextraktion mit Senfpulverlösung - Protokoll über drei Labor- und einen Freilandversuch. (unpublished)

Filser J & Krogh PH (2002) Interactions between *Enchytraeus crypticus*, collembolans, gamasid mites and Barley plants - a greenhouse experiment. Natura Jutlandica Occasional papers 2: 32-42

Finlay RD (1985) Interactions between soil micro-arthropods and endomycorrhizal associations of higher plants. In: Fitter, A. H., Atkinson, D. (Eds.), Biological interaction in soil (pp 319-331). Blackwell, Oxford

Fründ H-C & Jordan B (2004) Eignung verschiedener Senfzubereitungen als Alternative zu Formalin für die Austreibung von Regenwürmern. Mitteilungen der Deutschen Bodenkundlichen Gesellschaft 103: 25-26

Frouz J (2000) Changes in terrestrial chironomid (Diptera: Chironomidae) community after soil drainage. In: Hoffrichter, O. (Ed.), Late 20th Century Research on Chironomidae: an Anthology from the 13th International Symposium on Chironomidae (pp 285-289). Shaker Verlag, Aachen

Frouz J & Syrovátka O (1995) The effect of peat meadow drainage on soil dwelling dipteran communities - a preliminary report. Dipterologica Bohemoslovaca 7: 47-54

Frouz J (1999) Use of soil dwelling Diptera (Insecta, Diptera) as bioindicators: a review of ecological requirements and response to disturbance. Agriculture, Ecosystems and Environment 74: 167-186

Fukuhara H & Sakamoto M (1987) Enhancement of inorganic nitrogen and phosphate release from lake sediment by tubificid worms and chironomid larvae. Oikos 48: 312-320

Górny M (1984) Studies on the relationship between enchytraeids and earthworms. In: Szegi, J. (Ed.), Soil Biology and Conservation of the Biosphere, Vol. 1 + 2 (pp 769-778), Budapest

Giere O & Hauschildt D (1979) Experimental studies on the life-cycle and production of the littoral oligochaete *Lumbricillus lineatus* and its response to oil pollution. In: Naylor, E., Hartnall, R. G. (Eds.), Cyclic phenomena in marine plants and animals (pp 113-122). Pergamon, Oxford

Göbel C (2003) Bodenmakrofauna als Nahrungsgrundlage für Wiesenvögel: Auswirkungen auf Nistplatzwahl und Bruterfolg von Kiebitz, Uferschnepfe, Bekassine und Rotschenkel. Diploma thesis, University of Bremen, Bremen

Graefe U (1998) Annelidenzönosen nasser Böden und ihre Einordnung in Zersetzergesellschaften. Mitteilungen der Deutschen Bodenkundlichen Gesellschaft 88: 109-112

Graefe U (1987) Eine einfache Methode der Extraktion von Enchytraeiden aus Bodenproben. In: Koehler, H., Beck, L. (Eds.), Protokoll des Workshops zu Methoden der Mesofaunaerfassung und zu PCB-Wirkungen auf Collembolen und andere Mesofauna-Gruppen (pp 17). University of Bremen, Bremen

Graefe U & Schmelz RM (1999) Indicator values, strategy types and life forms of terrestrial Enchytraeidae and other microannelids. Newsletter on Enchytraeidae 6: 59-67

Graff O (1983) Unsere Regenwürmer. Schaper, Hannover

Grimm NB, Gergel SE, McDowell WH, Boyer EW, Dent CL, Groffman PM, Hart SC, Harvey JW, Johnston CA, Mayorga E, McClain ME & Pinay G (2003) Merging aquatic and terrestrial perspectives of nutrient biogeochemistry. Oecologia 442: 485-501

Gronstol GB, Solhoy T & Loyning MK (2000) A comparison of mustard, household detergent and formalin as vermifuges for earthworm sampling. Fauna Norvegica 20: 27-30

Gunn A (1992) The use of mustard to estimate earthworm populations. Pedobiologia 36: 65-67

Haimi J (1995) Effects of an introduced earthworm *Aporrectodea velox* on nutrient dynamics of forest soil. Acta Zoologica Fennica 196: 67-70

Haimi J & Huhta V (1990) Effects of earthworms on decomposition processes in raw humus forest soil: A microcosm study. Biology and Fertility of Soils 10: 178-183

Handke K (1993) Auswirkungen winterlicher Überstauungen auf die Fauna eines Grünland-Graben-Gebietes. Verhandlungen der Gesellschaft für Ökologie 22: 57-64

Handke K, Kundel W, Müller H-U, Riesner-Kabus M & Schreiber K-F (1999) Erfolgskontrolle zu Ausgleichs- und Ersatzmaßnahmen für das Güterverkehrszentrum Bremen in der Wesermarsch - 10 Jahre Begleituntersuchungen zu Grünlandextensivierung, Vernässung und Gewässerneuanlagen. Institut für Landschaftsökologie der Westfälischen Wilhelms-Universität, Münster

Hansen JD & Castelle AJ (1999) Insect diversity in soils of tidal and non-tidal wetlands of Spencer Island, Washington. Journal of the Kansas Entomol.Soc. 72: 262-272

Hartenstein R & Amico L (1983) Production and carrying capacity for the earthworm *Lumbricus terrestris* in culture. Soil Biology and Biochemistry 15: 51-54

Haslam SM, Klötzli F, Sukopp H & Szczepanski A (1998) The management of wetlands. In: Westlake, D. F., Kvet, J., Szczepanski, A. (Eds.), The production ecology of wetlands (pp 405-464). University Press, Cambridge

Haslam SM (2003) Understanding Wetlands - Fen, Bog and Marsh. Taylor and Francis, New York

Haynes RJ, Fraser PM, Piercy JE & Tregurtha RJ (2003) Casts of *Aporrectodea caliginosa* (Savigny) and *Lumbricus rubellus* (Hoffmeister) differ in microbial activity, nutrient availability and aggregate stability. Pedobiologia 47: 882-887

Healy B (1987) The depth distribution of Oligochaeta in an Irish quaking marsh. Hydrobiologia 155: 235-247

Hedlund K & Augustsson A (1995) Effects of enchytraeid grazing on fungal growth and respiration. Soil Biology and Biochemistry 27: 905-909

Hefting M (2003) Nitrogen transformation and retention in riparian buffer zones. PhD thesis, University of Utrecht, Utrecht

Hefting MM, Bobbink R & De Caluwe H (2003) Nitrous oxide emission and denitrification in chronically nitrate-loaded riparian buffer zones. Journal of Environmental Quality 32: 1194-1203

Helling B (1997) Einfluss der Regenwürmer auf die Stickstoff-Mineralisation und die bodenbiologische Aktivität landwirtschaftlich genutzter Flächen bei verschiedenen N-Düngern. Diss., Technical University Braunschweig, Braunschweig

Helling B & Kämmerer A (1998) Mehrjähriges Monitoring der Regenwurmfauna (Oligochaeta: Lumbricidae) extensiv genutzter Niedermoorböden im Drömling. Braunschweiger naturkundliche Schriften 5: 583-595

Hentze-Diesing W (1990) Auswirkungen von erhöhten Wasserständen in den Borgfelder Wümmewiesen - Literaturstudie unter besonderer Berücksichtigung bodenkundlicher Aspekte. Report for Umweltstiftung WWF-Deutschland, Projekt Wümmewiesen, Bremen

Hildebrandt J (1995) Anpassungen von Wirbellosen an Überschwemmungen und erhöhte Wasserstände. Berichte der Alfred-Töpfer-Akademie für Naturschutz (NNA) 2: 81-85

Hildebrandt J (1995) Erfassung von terrestrischen Wirbellosen in Feuchtgrünlandflächen im norddeutschen Raum - Kenntnisstand und Schutzkonzepte. Zeitschrift für Ökologie und Naturschutz 4: 181-201

Hindell RP, McKenzie BM, Tisdall JM & Silvapulle MJ (1994) Relationships between casts of geophagous earthworms (Lumbricidae, Oligochaeta) and matric potential. 1) Cast production, water-content, and bulk-density. Biology and Fertility of Soils 18: 119-126

Högger CH (1993) Mustard flour instead of formalin for the extraction of earthworms in the field. Bulletin BGS 17: 5-8

Höper H & Kleefisch B (2001) Untersuchung bodenbiologischer Parameter im Rahmen der Boden-Dauerbeobachtung in Niedersachsen: Bodenbiologische Referenzwerte und Zeitreihen. Niedersächsisches Landesamt für Bodenforschung NLfB, Hannover

Hoffmann CC, Kronvang B & Jacobsen JP (2004) Changes in nitrogen and phosphorus concentrations during flooding events in a riparian meadow. In: Verhoeven, J., Dorland, E. (Eds.), 7th Intecol International Wetlands Conference, Book of Abstracts (pp 133). University of Utrecht, Utrecht

Hoffmann CC, Kronvang B, Andersen I & Jacobsen JP (2003) Fate of nutrients during flooding events in a riparian wetland. In: Anonymous, Proceedings of the Warsaw conference of ECOFLOOD, Towards natural flood reduction strategies, International Conference Warsaw

Hofmeister F (2003) Valuation of ecosystem services - an intergrated dynamic approach (Der Wert von Feuchtgebieten aus Perspektive der Umweltökonomik). In: Society of Environmental Toxicology and Chemistry (Ed.), Tagungsband der 8. Jahrestagung der SETAC: "New Blood in Ecotoxicology" (pp 173-174). Heidelberg

International Organization for Standardization 2004, ISO CD 23611-1: Soil quality - sampling of soil invertebrates. Part 1: Hand-sorting and formalin extraction of earthworms: ISO Standards

Janhoff D (1992) Vegetationskundliche und standortökologische Untersuchungen im NSG "Borgfelder Wümmewiesen" - Grünland und Kleingewässer - Eine Studie zum Pflege- und Entwicklungsplan. Volume I - III. Planungsgruppe Grün, Report, WWF project Wümmewiesen, Bremen

Kalbitz K, Rupp H & Meissner R (2002) N-, P- and DOC-dynamics in soil and groundwater after restoration of intensively cultivated fens. In: Broll, G., Merbach, W., Pfeiffer, E.-M. (Eds.), Wetlands in Central Europe - Soil organisms, soil ecological processes and trace gas emissions Springer, Berlin

Kampichler C, Bruckner A, Baumgarten A, Berthold A & Zechmeister-Boltenstern S (1999) Field mesocosms for assessing biotic processes in soils: How to avoid side effects. European Journal of Soil Biology 35: 135-143

Kampichler C, Bruckner A & Kandeler E (2001) Use of enclosed model ecosystems in soil ecology: a bias towards laboratory research. Soil Biology and Biochemistry 33: 269-275

Karsten G (1997) Mikrobielle Populationen und Prozesse im Darm von Regenwürmern (Oligochaeta, Lumbricidae). Universität Bayreuth.

Keplin B & Broll G (2002) Earthworm coenoses in wet grassland of Northwest-Germany. Effects of restoration management on a histosol and a gleysol. In: Broll, G., Merbach, W., Pfeiffer, E.-M. (Eds.), Wetlands in Central Europe - Soil organisms, soil ecological processes and trace gas emissions (pp 11-34). Springer, Berlin

Keplin B, Hoffmann U & Broll G (1995) Extensivierung einer Pufferzone zum Schutz eines Hochmoorrestes - Auswirkungen der Wiedervernässung auf die Lumbricidenfauna. Mitteilungen der Deutschen Bodenkundlichen Gesellschaft 76: 819-822

Kleefisch B & Knes J (1997) Das Bodendauerbeobachtungsprogramm von Niedersachsen - Methoden und Ergebnisse. E. Schweizerbartsche Verlagsbuchhandlung, Stuttgart

Klok C, Zorn M, Koolhaas JE, Eijsackers H & Van Gestel CAM (2004) Maintaining viable earthworm populations in frequently inundated river floodplains. Does plasticity in maturation in *Lumbricus rubellus* promote population survival? In: Zorn, M. (Ed.), The floodplain upside down: Interactions between earthworm bioturbation, flooding and pollution. PhD thesis (pp 55-74). Vrije Universiteit, Amsterdam

Koerselmann W, Van Kerkhoven MB & Verhoeven J (1993) Release of inorganic N, P and K in peat soil: effect of temperature, water chemistry and water level. Biogeochemistry 20: 63-81

Krishnamoorthy RV (1990) Mineralization of phosphorus by faecal phosphatases of some earthworms of the Indian tropics. Proceedings of the Indian Academic Society (Animal Science) 99: 509-518

Lawrence AP & Bowers MA (2004) A test of "hot" mustard extraction method of sampling earthworms. Soil Biology and Biochemistry 34: 629-639

Le Bayon R-C & Binet F (1999) Rainfall effects on erosion of earthworm casts and phosphorus transfers by water runoff. Biology and Fertility of Soils 30: 7-13

Le Bayon R-C & Binet F (2001) Earthworm surface casts affect soil erosion by runoff water and phosphorus transfer in a temperate maize crop. Pedobiologia 45: 430-442

Lee KE (1985) Earthworms. Their ecology and relationships with soils and land use. Academic Press, Sydney

Lewandowski J & Hupfer M (2005) Effect of macrozoobenthos on two-dimensional small-scale heterogeneity of pore water phosphorus concentrations in lake sediments and phosphorus cycling in lakes: a laboratory study. Limnologia Oceanographica 50: 141-155

Ma W-C, Siepel H & Faber JH (1997) Bodemverontreiniging in de uiterwaarden: een bedreiging voor de terrestrische macroinvertebraten? Institute for Forestry and Nature Research (IBN-DLO), Wageningen, Lelystad

Mack AR (1963) Biological activity and mineralization of nitrogen in three soils as induced by freezing and drying. Canadian Journal of Soil Science 43: 316-324

Maculec G & Chmielewski K (1994) Earthworm communities and their role in hydrogenic soil. In: Anonymous, Proceedings of the international symposium: Conservation and management of fens (pp 417-427). Warsaw

Marinissen JCY & Van den Bosch F (1992) Colonization of new habitats by earthworms. Oecologia 91: 371-376

Marinissen JCY & de Ruiter PC (1993) Contribution of earthworms to carbon and nitrogen cycling in agro-ecosystems. Agriculture, Ecosystems and Environment 47: 59-74

Martin A & Marinissen JCY (1993) Biological and physicochemical processes in excrements of soil animals. Geoderma 56: 331-347

Mather JG & Christensen O (1992) Surface migration of earthworms in grassland. Pedobiologia 36: 51-57

Mather JG & Christensen B (1988) Surface movements of earthworms in agricultural land. Pedobiologia 32: 399-405

McClain ME, Boyer EW, Dent CL, Gergel SE, Grimm NB, Groffmann PM, Hart SC, Harvey JW, Johnston CA, Mayorga E, McDowell WH & Pinay G (2003) Biogeochemical hot spots and hot moments at the interface of terrestrial and aquatic ecosystems. Ecosystems 6: 301-312

McInerney M & Bolger T (2000) Temperature, wetting cycles and soil texture effects on carbon and nitrogen dynamics in stabilized earthworm casts. Soil Biology and Biochemistry 32: 335-349

Meenken GA (1999) Zum Wiesenvogelschutz im Bruthabitat der Strohauser Plate - Nahrungsangebot in Abhängigkeit von Stocherfähigkeit und Wasserverhältnissen. Diploma thesis. Hochschule Vechta.

Meuleman AFM (1999) Performance of treatment wetlands. PhD thesis, University of Utrecht, Utrecht

Mitsch WJ & Gosselink JG (1993) Wetlands. Van Nostrand Reinhold, New York

Nagel R (1996) Die Bedeutung von Regenwürmern für den C- und N-Umsatz in einer heterogenen Agrarlandschaft. Technical University München, München

Nagel RF, Fromm H & Beese F (1995) The influence of earthworms and soil mesofauna on the C and N mineralization in agricultural soils - a microcosms study. Acta Zoologica Fennica 196: 22-26

Nagel RF & Beese F (1992) Veränderung des Transportverhaltens gelöster Stoffe durch Regenwurmgänge. Mitteilungen der Deutschen Bodenkundlichen Gesellschaft 67: 107-110

Nielsen CO & Christensen B (1959) The Enchytraeidae - Critical revision and taxonomy of European species. Aarhus

Obrdlik P, Falkner G & Castella E (1995) Biodiversity of Gastropoda in European floodplains. Archiv für Hydrobiologie, Supplementband 101: 339-356

Parkin TB & Berry EC (1994) Nitrogen transformations associated with earthworm casts. Soil Biology and Biochemistry 26: 1233-1238

Parle JN (1963) A microbiological study of earthworm casts. Journal of Geneal Microbiology 31: 13-22

Paul EA & Clark FE (2004) Soil microbiology and biochemistry. Academic Press, New York

Petersen H & Luxton M (1982) A comparative analysis of soil fauna populations and their role in decomposition processes. Oikos 39(3): 290-388

Pizl V (1999) Earthworm communities in hardwood floodplain forests of the Morava and Dyje rivers as influenced by different inundation regimes. Ekológia Bratislava 18, Supplement: 197-204

Pizl V & Tajovský K (1998) Vliv letní povodne na pudni makrofaunu luzniho lesa v litovelskem pomoravi (Changes of terrestrial invertebrate communities after disastrous summer flood in 1997 in four ecosystems in the Central Moravian floodplain, Czech Rebublic), Prague (unpublished)

Plum NM & Filser J (2005) Floods and drought: Response of earthworms and potworms (Oligochaeta: Lumbricidae, Enchytraeidae) to hydrological extremes in wet grassland. Pedobiologia 49(3): 443-453

Plum NM (2005) Terrestrial invertebrates in flooded grassland - a literature review. Wetlands 25: 721-737

Priesner E (1961) Nahrungswahl und Nahrungsverarbeitung bei der Larve von *Tipula maxima*. Pedobiologia 1: 25-37

Raw F (1959) Estimating earthworm populations by using formalin. Nature 184: 1661-1662

Reddy KR, Patrick Jr WH & Philips RE (1980) Evaluation of selected processes controlling nitrogen loss in a flooded soil. Soil Society of America Journal 44: 1241-1246

Reimer G & Zulka P (1994) Ökologische Auswirkungen von Überflutungen auf die Fischfauna der March. Wissenschaftliche Mitteilungen des Niederösterreichischen Landesmuseums 8: 191-201

Richardson CJ & Schlesinger WH (2004) Biogeochemistry in wetlands: A global perspective. In: Verhoeven, J., Dorland, E. (Eds.), 7th Intecol International Wetlands Conference, Book of Abstracts (pp 258). University of Utrecht, Utrecht

Richardson CJ & Marshall PE (1986) Processes controlling movement, storage, and export of phosphorus in a fen peatland. Ecological Monographs 56: 279-302

Richardson CJ (1999) The role of wetlands in storage, release, and cycling of phosphorus on the landscape: a 25 year retrospective. In: Reddy, K. R. (Ed.), Phosphorus Biogeochemistry in Sub-Tropical Ecosystems CRC Press/Lewis Publishers, Florida

Römbke J, Beck L, Dreher P, Hund-Rinke K, Jänsch S, Kratz W, Pieper S, Ruf A, Spelda J & Woas S (2002) Entwicklung von bodenbiologischen Bodengüteklassen für Acker- und Grünlandstandorte. Umweltbundesamt, Bonn

Rodieck B, Schriefer T, Beckmann M, Timmermann U & Gefken T (1992) Bodenökologische Untersuchung in den Borgfelder Wümmewiesen. Report for Umweltstiftung WWF Deutschland, Bremen

Roots BI (1956) The water relations of earthworms. II. Resistance to dessication and immersion and behaviour when submerged and when allowed a choice of environment. Journal of Experimental Biology 33: 29-44

Rusek J (1984) Zur Bodenfauna in drei Typen von Überschwemmungswiese in Süd-Mähren. Rozpravy Ceskoslovenské Akademie Ved 94: 1-126

Ruz-Jerez BE, Ball PR & Tillman RW (1992) Laboratory assessment of nutrient release from a pasture soil receiving grass or clover residues, in the presence or absence of *Lumbricus rubellus* or *Eisenia foetida*. Soil Biology and Biochemistry 24: 1529-1534

Satchell JE & Martin K (1984) Phosphatase activity in earthworms faeces. Soil Biology and Biochemistry 16: 191-194

Schaefer M & Schauermann J (1990) The soil fauna of beech forests: comparison between a mull and a moder soil. Pedobiologia 34: 299-314

Scheffer,B & Ausborn,R, Anonymous,1998, Uferstreifen an künstlichen Gewässern und umweltgerechte Landwirtschaft in ihrer Auswirkung auf künstliche Gewässer im Lande Bremen - Kurzfassung. Bericht des Niedersächsischen Landesamtes für Bodenforschung (NLfB), Bodentechnologisches Institut: Bremen: Senator für Frauen, Gesundheit, Jugend, Soziales und Umweltschutz der Freien Hansestadt.

Schekkerman H (1997) Graslandbeheer en groeimogelijkheden voor weidevogels. Dienst landelijk gebied, Institut voor Bos- en Natuurbeheer IBN-DLO, Wageningen

Scheu S (1987a) Microbial activity and nutrient dynamics in earthworm casts (Lumbricidae). Biology and Fertility of Soils 5: 230-234

Scheu S (1987b) The role of substrate feeding earthworms (Lumbricidae) for bioturbation in a beechwood soil. Oecologia 72: 192-196

Scheu S (1993) There is an earthworm mobilizable nitrogen pool in soil. Pedobiologia 37: 243-249

Scheu S (1995) Mixing of litter and soil by earthworms: effects on carbon and nitrogen dynamics - a laboratory experiment. Acta Zoologica Fennica 196: 33-40

Schindler-Wessels ML, Bohlen PJ, McCartney DA, Subler S & Edwards CA (1997) Earthworm effects on soil respiration in corn agroecosystems receiving different nutrient inputs. Soil Biology and Biochemistry 29: 409-412

Schmidt O (2001) Time-limited soil sorting for long-term monitoring of earthworm populations. Pedobiologia 45: 69-83

Schrader S, Langmaack M & Helming K (1997) Impact of Collembola and Enchytraeidae on soil surface roughness and properties. Biology and Fertility of Soils 25: 396-400

Schrader S (1994) Influence of earthworms on the pH conditions of their environment by cutaneous mucus secretion. Zoologischer Anzeiger 233: 211-219

Schröder F (1980) Bestandsaufnahme der Molluskenfauna im Bereich des östlichen Hollerlandes, Bremen. (unpublished)

Schultz J (2000) Handbuch der Ökozonen. Ulmer UTB, Stuttgart

Schwert DP & Dance KW (1979) Earthworm cocoons as a drift component in a southern Ontario stream. Can Field-Nat 93: 180-183

Senator für Bau, Umwelt und Verkehr (2004) Umsetzung der EG-Wasserrahmenrichtlinie im Land Bremen: detaillierte Beschreibung der Gewässer mit Einzugsgebieten > 10 km²: Freie Hansestadt Bremen, Bremen

Setälä H, Martikainen E, Tyynismaa M & Huhta V (1990) Effects of soil fauna on leaching of nitrogen and phosphorus from experimental systems simulating coniferous forest floor. Biology and Fertility of Soils 10: 170-177

Sharpley AN, Syers JK & Springett JA (1979) Effect of surface-casting earthworms on the transport of phosphorous and nitrogen in surface runoff from pasture. Soil Biology and Biochemistry 11: 459-462

Sharpley AN & Syers JK (1976) Potential role of earthworm casts for the phosphorus enrichment of runoff waters. Soil Biology and Biochemistry 8: 341-346

Sharpley AN & Syers JK (1977) Seasonal variation in casting activity and in the amounts and release to solution of phosphorus in earthworm casts. Soil Biology and Biochemistry 9: 227-231

Shipitalo MJ, Protz R & Tomlin AD (1988) Effect of the diet on the feeding and casting activity of *Lumbricus terrestris* and *L. rubellus* in laboratory culture. Soil Biology and Biochemistry 20: 233-237

Simek M & Pizl V (1991) The effect of earthworms (Lumbricidae) on nitrogenase activity in the soil. Biology and Fertility of Soils 7: 370-373

Simpson RL, Whigham DF & Walker R (1978) Seasonal patterns of nutrient movement in a freshwater tidal marsh. In: Good, R. E., Whigham, D. F., Simpson, R. L. (Eds.), Freshwater wetlands (pp 243-258). Academic Press, New York

Sims GK & Gerard BM (1985) Earthworms: Synopses of the British Fauna 31. Brill, Leiden

Spang WD (1996) Die Eignung von Regenwürmern (Lumbricidae), Schnecken (Gastropoda) und Laufkäfern (Carabidae) als Indikatoren für auentypische Standortbedingungen - Eine Untersuchung im Oberrheintal.

Springett JA, Brittain JE & Springett BP (1970) Vertical movement of Enchytraeidae (Oligochaeta) in moorland soils. Oikos 21: 21 pp.

Stockdill SMJ (1982) Effect of introduced earthworms on the productivity of New Zealand pastures. Pedobiologia 24: 29-35

Striganova BR, Pantoshderimova TD & Tiunov AV (1993) Comparative estimation of nitrogen fixation activity in the intestine of various species of earthworms. Izvestiya Akademii Nauk, Serija Biologiceskaja 2: 257-263

Subler S, Baranski CM & Edwards CA (1997) Earthworm additions increased short-term nitrogen availability and leaching in two grain-crop agroecosystems. Soil Biology and Biochemistry 29: 413-421

Subler S, Parmelee RW & Allen MF (1998) Earthworm and nitrogen mineralization in corn agroecosystems with different nutrient amendments. Applied Soil Ecology 9: 295-301

Sulkava P, Huhta V & Laakso J (1996) Impact of soil faunal structure on decomposition and N-mineralisation in relation to temperature and moisture in forest soil. Pedobiologia 40: 505-513

Svensson BH, Bostrom U & Klemedtsson L (1986) Potential for higher rates of denitrification in earthworm casts than in the surrounding environment. Biology and Fertility of Soils 2: 147-149

Syers JK, Sharpley AN & Keeney DR (1979) Cycling of nitrogen by surface casting earthworms in a pasture ecosystem. Soil Biology and Biochemistry 11: 181-185

Tabeling H & Düttmann H (2002) Einfluss von Düngung und Überstauung auf die Bodenmegafauna von Niedermoorgrünländern der südlichen Dümmerniederung (Niedersachsen). In: Düttmann, H., Ehrnsberger, R., Faida, I. (Eds.), Wiesenvogelschutz in Norddeutschland und den Niederlanden, Symposium an der Hochschule Vechta, Abstracts (pp 63-64), Vechta

Tajovský K (1998) Diversity of terrestrial isopods (Oniscidea) in flooded and nonflooded ecosystems of southern Moravia, Czech Republic. Israel Journal of Zoology 44: 311-322

Tezuka Y (1990) Bacterial regeneration of ammonium and phosphate as affected by carbon:nitrogen:phosphorus ratio of organic substrates. Microbial Ecology 19: 227-238

Thielemann U (1986) Elektrischer Regenwurmfang mit der Oktett-Methode. Pedobiologia 29: 269-302

Tiefenbrunner A & Tiefenbrunner W (1993) Zwei chemische Lumbriciden-Extraktionsmethoden im Vergleich: Senf und Formalin. Pflanzenschutzberichte 53: 133-140

Tiwari SC & Mishra RR (1993) Fungal abundance and diversity in earthworm casts and in uningested soil. Biology and Fertility of Soils 2: 131-134

Topp W (1981) Biologie der Bodenorganismen. UTB Quelle und Meyer, Heidelberg

Trepel M & Kluge W (2002) Analyse von Wasserpfaden und Stofftransformationen in Feuchtgebieten zur Bewertung der diffusen Austräge. KA - Wasserwirtschaft, Abwasser, Abfall 49: 807-815

Updegraff K, Pastor J, Bridgham SD & Johnston CA (1995) Environmental and substrate controls over carbon and nitrogen mineralization in northern wetlands. Ecological Applications 5: 151-163

Volz P (1976) Die Regenwurm-Population im Naturschutzgebiet Hördter Rheinaue und ihre Abhängigkeit vom Feuchtigkeitsregime des Standortes. Mitteilungen der Pollichia 64: 110-120

Wallwork JA (1983) Earthworm biology. Edward Arnold Publishers Ltd., London

Weiß B (1994) Einfluss von Regenwürmern auf ausgewählte Enzymaktivitäten und den Stickstoff-Haushalt in landwirtschaftlich genutzten Böden. Technical University Braunschweig, Braunschweig

Weigmann G & Schumann M (1999) Bodentypen und Schwermetallbelastung von Böden, Pflanzen und Bodentieren in Überschwemmungsgebieten des Unteren Odertals. In: Dohle, W., Bornkamm, R., Weigmann, G. (Eds.), Das Untere

Odertal: Auswirkungen der periodischen Überschwemmungen auf Biozönosen und Arten (pp 23-38). E. Schweizerbart'sche Verlagsbuchhandlung, Stuttgart

Weigmann G & Wohlgemuth-von Reiche D (1999) Vergleichende Betrachtungen zu den Überlebensstrategien von Bodentieren. In: Dohle, W., Bornkamm, R., Weigmann, G. (Eds.), Das Untere Odertal: Auswirkungen der periodischen Überschwemmungen auf Biozönosen und Arten (pp 229-240). E. Schweizerbart'sche Verlagsbuchhandlung, Stuttgart

Well R, Augustin J & Meyer K (2002) In situ measurement of denitrification and N_2O production in the saturated zone of three Eutric Histosols and of a Mollic Gleysol. In: Broll, G., Merbach, W., Pfeiffer, E.-M. (Eds.), Wetlands in Central Europe (pp 172-176). Springer, Berlin

Wienk LD, Verhoeven JTA, Coops H & Portielje R (2000) Peilbeheer en nutrienten. Report for Ministerie van Verkeer en Waterstaat, Directoraat-Generaal Rijkswaterstaat, Lelystad

Williams BL & Griffiths BS (1989) Enhanced nutrient mineralization and leaching from decomposing sitka spruce litter by enchytraeids worms. Soil Biology and Biochemistry 21: 183-188

Wolters V (1988) Effects of *Mesenchytraeus glandulosus* (Oligochaeta, Enchytraeidae) on decomposition processes. Pedobiologia 32: 387-398

Zaborski ER (2003) Allyl isothiocyanate: an alternative chemical expellent for sampling earthworms. Applied Soil Ecology 22: 87-95

Zachariae G (1967) Die Streuzersetzung im Köhlgartengebiet. In: Graff, O., Satchell, J. E. (Eds.), Progress in Soil Biology (pp 490-506). Braunschweig, Amsterdam

Zak D, Gelbrecht J & Steinberg CEW (2004) Phosphorus retention at the redox interface of peatlands adjacent to surface waters in northeast Germany. Biogeochemistry 70: 359-370

Zerm M (1997) Distribution and phenology of *Lamyctes fulvicornis* and other lithobiomorph centipedes in the floodplain of the Lower Oder Valley, Germany (Chilopoda, Henicopidae: Lithobiidae). Entomologia Scandinavica Supplements 51: 125-132

Zerm M (1999) Vorkommen und Verteilung von Tausendfüßern, Hundertfüßern, Zwergfüßern (Myriapoda: Diplopoda, Chilopoda, Symphyla) und Landasseln (Isopoda: Oniscidea) in den Auen des Unteren Odertals. In: Dohle, W., Bornkamm, R., Weigmann, G. (Eds.), Das Untere Odertal: Auswirkungen der periodischen Überschwemmungen auf Biozönosen und Arten (pp 197-210). E. Schweizerbart'sche Verlagsbuchhandlung, Stuttgart

Zöfel P (1988) Statistik in der Praxis. UTB Gustav Fischer, Stuttgart

Zorn M, Van Gestel CAM, Morrien E, Wagenaar M & Eijsackers H (2004a) Flooding responses of *Allolobophora chlorotica*, *Aporrectodea caliginosa* and *Lumbricus rubellus* in a laboratory-controlled environment. In: Zorn, M. (Ed.), The floodplain upside down: Interactions between earthworm bioturbation, flooding and pollution (pp 39-54). Free University (VU), Amsterdam

Zorn M, Van Gestel CAM & Eijsackers H (2004b) Species-specific earthworm population responses in relation to flooding dynamics in a Dutch floodplain soil. Pedobiologia 43: 189-198

Zulka KP (1991) Überflutung als ökologischer Faktor: Verteilung, Phänologie und Anpassung der Diplopoda, Lithobiomorpha und Isopoda in den Flussauen der March.
PhD thesis, University of Wien.

Chapter 12 - References

Annexe

Seasonal dynamics of terrestrial and aquatic invertebrate bycatch in flooded grassland

Next to earthworms and enchytraeids, various groups of terrestrial and aquatic invertebrates were sampled during field investigations (Chapter 5). Terrestrial macroinvertebrates were found as bycatch of the mustard extraction of earthworms and the following hand sorting. With the wet funnel method for enchytraeids, other representatives of soil mesofauna were occasionally sampled. Additionally, aquatic organisms living in pore water of floodwater were observed.

As no standard methods were used, the data are not representative. Terrestrial and aquatic families within a group (e.g., Diptera larvae and Nematoda) could not always be distinguished properly. However, the data can be compared across sites and time, giving a coarse impression of the dynamics these groups undergo due to the hydrological extremes in the study sites.

Endogeic and epigeic macrofauna

Endogeic insect larvae were present in all three sites (Fig. A.1). Amongst the macrofauna found by hand sorting, Bibionidae (Diptera) and Curculionidae (Coleoptera) were the most abundant families. Ceratopogonidae and Chironomidae were the most numerous families found with the wet funnel technique. They occurred in high abundances in the wet year 2002 (maximum value of nearly 8000 ind. m^{-2} in the peat soil in September 2002). Gastropods played an important role in the two sites in Bremen, especially in summer 2002, even after the Wümme flood on the peat soil, while they were completely absent in the Gorleben site. Mites (Acari: Oribatida, Gamasina) were abundant in most samplings of the marsh and the gley soil, whereas in the peat they only appeared with the last sampling in September 2003. In the Ochtum site, spiders and big red epigeic mites were abundant in autumn of both years, with a dominance of Gastropods (*Succinea* spec. and others) in 2002 and of adult insects (most of them Cicadina) in 2003. Collembola were always absent after the flood events and occurred regularly in autumn samplings.

Mesofauna and aquatic organisms

In all three soils, the most abundant and regular bycatch of the wet funnel extraction of enchytraeids were nematodes (Fig. A.2). Only free-living individuals that could be seen and sorted at 12-fold magnification were counted, disregarding very small individuals. The highest nematode abundance was found in the marsh soil, where a strong decline during the early dry summer of 2003 could be observed, followed by a higher abundance in July. In contrast, maximum values in the gley soil were observed after summer flood. Summer drought in 2003 obviously lead to a total breakdown of the complete soil fauna, even of nematode populations. Small dipteran and other insect larvae were also present in most samplings on all three soils, most abundant in the wet year 2002 (maximum value of nearly 8000 ind. m^{-2} in the peat soil in September 2002, Fig. A.3). A hemiedaphic collembolan species (probably *Isotoma notabilis*) had a mass occurrence before the summer flood in the Gorleben site. Collembola were always absent after the flood events and occurred regularly in autumn samplings. Mites (Acari: Oribatida, Gamasina) were abundant in most samplings of the marsh and the gley soil, whereas in the peat, they only appeared with the last sampling in September 2003.

Aquatic organisms

At normal or high soil humidity, aquatic organisms (i.e. *Lumbriculus variegatus*, Tubificidae: *Rhyacodrilus* spec., Copepoda, Ostracoda, *Daphnia* spec., aquatic dipteran larvae, Turbellaria and Rotatoria) were practically omnipresent. Maximum abundances were found after flood events and in autumn 2002. Only during the severe summer drought in 2003, their numbers were locally reduced to zero (June 2002, June/July 2003 in the gley soil; absence of lumbriculids in the peat soil in July 2003; Fig. A.4).

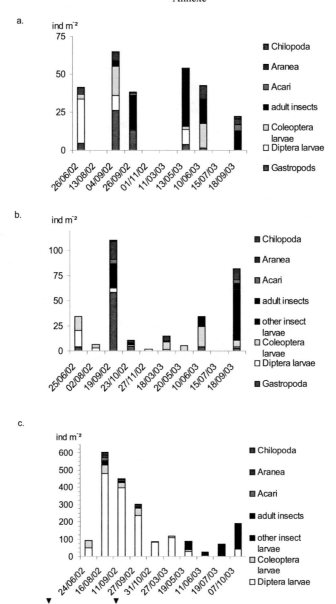

Figure A.1: Endogeic and epigeic macrofauna bycatch in earthworm samplings.
a. marsh soil, b. peat soil, c. gley soil. Arrows indicate flooding events (cf. 5.2.2)

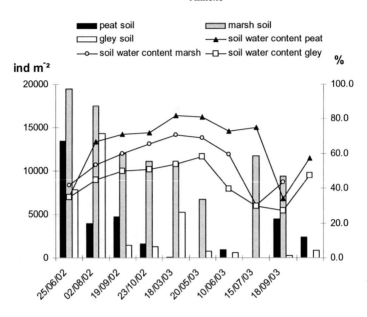

Figure A.2: Nematoda in enchytraeid samplings

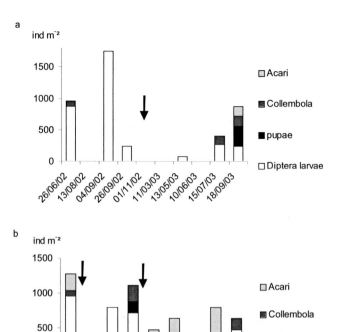

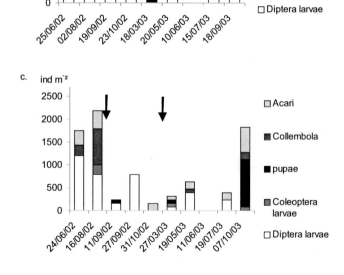

Figure A.3: Soil mesofauna bycatch in enchytraeid samplings
a. marsh soil, b. peat soil, c. gley soil. Arrows indicate flooding events (cf. 5.2.2)

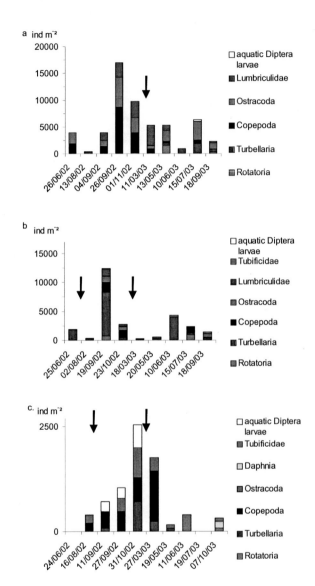

Figure A.4: Aquatic invertebrates in enchytraeid sampling
a. marsh soil, b. peat soil, c. gley soil. Arrows indicate flooding events (cf. 5.2.2)

List of publications

Plum, N.M. (2005): Terrestrial invertebrates in flooded grassland - a literature review
Wetlands 25 (3) 721-737

Plum, N.M., Filser, J. (2005): Floods and drought: Response of earthworms and potworms (Oligochaeta: Lumbricidae, Enchytraeidae) to hydrological extremes in wet grassland. Pedobiologia 49 (3)

Plum, N.M., Filser, J.: Earthworms and potworms (Lumbricidae, Enchytraeidae) mobilise soil nutrients from flooded grassland soils. Ecological Engineering (under revision)

Plum, N.M., Filser, J.: Temporal variation in nutrient release from a flooded peat soil depending on earthworms and enchytraeid density. European Journal of Soil Biology (under revision)

Plum, N.M: Effects of living and dead earthworms (Lumbricidae) on nutrient mobilisation dynamics from a flooded peat soil. Biogeochemistry (under revision)

Plum, N.M.: Worms and wetland water: the role of lumbricids and enchytraeids in nutrient mobilization from flooded soils. Abhandlungen und Berichte des Naturkundemuseums Görlitz, Special Issue on the Symposium "Floodplains: Hydrology, Soils, Fauna, and their Interactions" at the State Museum of Natural History, Görlitz, Germany, September 2005 (under revision)

Klok, C., Plum, N.M.: Earthworms in wetlands, do they adapt to flooding? Support from field data. Abhandlungen und Berichte des Naturkundemuseums Görlitz, Special Issue on the Symposium "Floodplains: Hydrology, Soils, Fauna, and their Interactions" at the State Museum of Natural History, Görlitz, Germany, September 2005 (under revision)

Annexe

Conference contributions

Plum, N.: Die Bedeutung von Regenwürmern und Kleinringelwürmern für den terrestrisch-aquatische Austauschprozesse in Feuchtgrünlandökosystemen: Ein Forschungskonzept. Oral presentation at the 18. Jahrestreffen der AG Bodenmesofauna at the University of Darmstadt, Germany, April 2002

Plum, N.: Die Bedeutung von Regenwürmern und Kleinringelwürmern für terrestrisch-aquatische Austauschprozesse in Feuchtgrünlandökosystemen. Poster at the Fachtagung "Wasserrahmenrichtlinie (WRRL) und Naturschutz at the NNA (Alfred Töpfer Akademie für Naturschutz) in Schneverdingen, Lüneburger Heide, Germany, October 2002

Plum, N.: Die Bedeutung von Regenwürmern und Kleinringelwürmern für terrestrisch-aquatische Austauschprozesse in Feuchtgrünlandökosystemen: Erste Ergebnisse zur Populationsdynamik im Freiland und erweitertes Forschungskonzept. Oral presentation at the 19. Jahrestreffen der AG Bodenmesofauna at the University of Soil Science (BOKU) in Vienna, Austria, April 2003

Plum, N.: Bodenmakrofauna und Enchytraeiden in überflutetem Grünland: Strategien und Schicksale von Grenzgängern zwischen Land und Wasser- Ergebnisse einer Literaturstudie. 20. Jahrestreffen der AG Bodenmesofauna at the State Museum of Natural History, Görlitz, Germany, April 2004

Plum, N.M., Filser, J.: Annelid worms in flooded grassland mobilize soil nutrients - Results of two laboratory experiments. Oral presentation at the 7th INTECOL International Wetlands Conference at the University of Utrecht, The Netherlands, July 2004

Plum, N.M., Filser, J.: Nutrient release from flooded peat soil as affected by earthworms and enchytraeid density. Oral presentation at the 34th Annual Conference of the German Ecological Society (GfÖ) at the Justus-Liebig-University in Gießen, Germany, August 2004

Plum, N.M., Filser, J.: Effects of earthworms and enchytraeids on nutrient mobilisation after flooding of wetland soils - A laboratory experiment. Oral presentation (by Juliane Filser) in Rouen, France, July 2004

Plum, N.M.: Worms and wetland water: the role of lumbricids and enchytraeids in nutrient mobilization from flooded soils. Oral presentation at the International Symposium "Floodplains: Hydrology, Soils, Fauna, and their Interactions" at the State Museum of Natural History, Görlitz, Germany, September 2005

Curriculum vitae/ Lebenslauf

Curriculum vitae

Nathalie Madeleine Plum, born on January 29th 1974 in
Heerlen (The Netherlands) studied Geography, Biology a
International Cooperation at the Technical University of
Aachen (RWTH) and at the University of Amsterdam
and got her Master of Arts (M.A.) in January 2000.
As a student assistant, she participated in the completion
of the textbook „Die Ökozonen der Erde" (The ecozones
the world, Prof. J. Schultz). Her master thesis was about
„Vegetation dynamics on coal mine heaps in the region o
Aachen-Erkelenz".

From February 2000 to November 2001 she worked at the Moor Research Station of the
University of Liège (Belgium) as a botanical cartographer in the European Interreg II-
project „Protection and management of border-crossing valleys and creeks in the
German-Belgian Nature Park High Fen-Eifel" as well as for the Wallonian government
in the framework of the European nature reserve network „Natura 2000".

The main work for the present thesis was performed during her time as scientific
assistant at the University of Bremen in the Center for Environmental Research and
Technology (UFT), Department of General and Theoretical Ecology from December
2001 to November 2004.

Since February 2005, she is in charge of the regional office of the NGO NABU
(Naturschutzbund Deutschland, member of BirdLife International) in the southern Pfalz
in Landau.

Lebenslauf

Nathalie Madeleine Plum, geboren am 29. Januar 1974 in Heerlen (Niederlande)
studierte Geografie, Biologie und Internationale Technische und Wirtschaftliche
Zusammenarbeit (ITWZ) an der Rheinisch-Westfälischen Technischen Hochschule
(RWTH) in Aachen und an der Universität von Amsterdam. Im Januar 2000 erlangte sie
den Magistergrad (Magistra Artium, M.A.) mit einer Arbeit über die
„Vegetationsdynamik auf Steinkohlenbergehalden im Aachen-Erkelenzer Revier". Als

studentische Hilfskraft war sie eingebunden in die Fertigstellung des Lehrbuchs „Die Ökozonen der Erde" (Prof. J. Schultz).

Von Februar 2000 bis November 2001 arbeitete sie als botanische Kartiererin in der Moorforschungsstation der Universität Lüttich (Belgien) im Europäischen Interreg II-Projekt „Schutz und Pflege grenzüberschreitender Täler und Wasserläufe im Deutsch-Belgischen Naturpark Hohes Venn-Eifel" sowie für die Wallonische Regierung im Rahmen des europäischen Schutzgebietsnetzwerks „Natura 2000".

Ein Großteil der Arbeit für die vorliegende Dissertation wurde während ihrer Zeit als Wissenschaftliche Mitarbeitern an der Universität Bremen im Zentrum für Umweltforschung und Umwelttechnologie (UFT), Abteilung Allgemeine und Theoretische Ökologie von Dezember 2001 bis November 2004 durchgeführt.

Seit Februar leitet sie die Regionalstelle Südpfalz des Naturschutzbund Deutschland NABU in Landau.

Das Manuskript wurde nach dem Dissertationskolloquium auf Grundlage der Gutachten leicht überarbeitet.

Versicherung

Die Autorin bestätigt, die vorliegende Arbeit ohne unerlaubte fremde Hilfe angefertigt, keine als die angegebenen Hilfsmittel benutzt und die den benutzten Werken wörtlich oder inhaltlich entnommenen Stellen als solche kenntlich gemacht zu haben.

Bremen, den 9.9.2005

Danksagung

Zuallererst möchte ich mich bei meiner Betreuerin Juliane Filser bedanken, die mir auch im überfülltesten Terminkalender stets eine Stunde einzuräumen wusste und mich „doktormütterlich" durch meine Krisen mit der Arbeit, dem Labor und der Wissenschaft an sich sowie der norddeutschen Tiefebene im Speziellen begleitet hat. Meiner gesamten Prüfungskommission (neben Juliane Filser: Michael Schirmer, Martin Diekmann, Broder Breckling, Enken Hassold und Jana Seeger) danke ich für die gute Bewertung der Arbeit.

Außerdem danke ich allen Kollegen der Allgemeinen und Theoretischen Ökologie, im IFÖE und im UFT für ihre Unterstützung bei der Vorbereitung der Arbeit und bei den Laborarbeiten. Besonders hervorzuheben sind die, die sich Zeit genommen haben, mich in Methoden einzuarbeiten (wenn nicht anders erwähnt, gehören sie der Arbeitsgruppe Allgemeine und Theoretische Ökologie an):

Marion Ahlbrecht (Vegetationsökologie und Naturschutzbiologie) in die Wasseranalytik am Fließinjektionsanalysator, auch wenn diese nie brauchbare Ergebnisse brachte,

Bernd Steinweg und Jörg Bierwirth (Aquatische Ökologie) in den Hach-Schnelltest,

Hartmut Koehler in die Extraktion von Enchyträen,

Elke Obert (IUV) in die Dr. Lange-Schnelltests,

Maike Schaefer in die Bestimmung von Regenwürmern,

Ute Uebers in die Laborarbeit im Allgemeinen, ins Binokulieren, Mikroskopieren und in die Bestimmung der Maximalen Bodenwasserhaltekapazität;

Michael Birkner (UFT) für die (teils kurzfristig notwendig gewordene) Wartung des Zahnkranzbohrers und die Herstellung von bodenlosen Eimern für das Freilandexperiment;

Holger Mebes für die Einführung in statistische Verfahren;

Falko Berger (UFT) für diverse spontane Hilfe mit Hardware-, Software- und Netzwerkproblemen, insbesondere für die pietätvolle Begleitung bei der notwendigen Einschläferung des alten Ecotox-Rechners, die dank seiner fachkundigen Handhabung vollkommen ohne Datenverlust vonstatten gehen konnte.

Danke auch an meinen Diplomanden Christian Göbel sowie an die PraktikantInnen Annika, Natalie Lürßen, Mareile, Svea, Michael Verhülsdonk, Aljosha Born, Lukas Reuter und Elias Schwab für die tatkräftige Unterstützung bei Labor- und Geländearbeiten. Eine besondere Stellung unter den Praktikanten hatten Christopher Klar, der treu und engagiert in den Schulferien immer wieder half, sowie Peer Cyriacks (BUND), der gute, aber dennoch fruchtlose Ideen zum Design des Freilandexperiments lieferte und mich mit seinen Hinweisen auf Rotmilan und Silberreiher aufmunterte, während mein entsetzter Blick bei den auslaufenden Mesokosmen verharrte.

Den Kollegen vom Senator für Bau, Umwelt und Verkehr (Henrich Klugkist, Andreas Nageler) danke ich für die Genehmigung zur Probenahme in den Naturschutzgebieten Borgfelder Wümmewiesen und Ochtumniederung bei Brokhuchting, ebenso den Kollegen vom NLfB für die Genehmigung zu den Freilandarbeiten auf der Bodendauer-beobachtungsfläche in Gorleben sowie die Bereitstellung von Datenmaterial zur Fläche. Insbesondere danke ich Bernd Kleefisch, Hubert Groh und Silke Hillebrand für das gemeinsame Wochenend-Abenteuer in Lüchow und Gorleben während der Elbeflut im August 2002. Gunnar Oertel vom WWF und Birgit Olbrich vom BUND sowie ihren Mitarbeitern gilt mein Dank ebenfalls für die Genehmigungen sowie für den regen fachlichen Austausch über die Flächen. Wolfhart Koehler, dem Pächter der Probefläche in der Ochtumniederung, danke ich herzlich für die Erlaubnis zur Durchführung des Freilandexperiments. Einem unbekannten Landwirt am Großen Moordamm entlang der Borgfelder Wümmewiesen danke ich für die Spende von Gefrierbeuteln zum Transport von Bohrkernen.

Mein besonderer Dank gilt Ulfert Graefe (Institut für Angewandte Bodenbiologie IfAB Hamburg) für die Bestimmung einiger Regenwürmer sowie für die Einführung in die Bestimmung von Enchyträen, vor allem für seinen behutsamen Versuch, in mir eine Faszination für diese kleinen unscheinbaren Tiere zu wecken. Auch seiner Frau Monika danke ich für die köstlichen mediterranen Imbisse und die erquickenden Gespräche über Literatur.

Ulrich Burkhard und Kerstin Mölter danke ich für die kollegiale Nachbarschaft in unserem Labor-Büro. Ulrich: diesmal *Omnia mea mecum porto*, hoffe ich, oder habt ihr noch Überreste meiner Besitztümer in den Schubladen gefunden?

Michael Schirmer (Aquatische Ökologie) und Martin Diekmann (Vegetationsökologie und Naturschutzbiologie) sowie diversen anderen Kollegen danke ich für die gute Zusammenarbeit in der Lehre. Dem Arbeitskreis Theorie im IFÖE, besonders Hauke Reuter und Broder Breckling, möchte ich danken für die vielen interessanten Diskussionen, die mein ökologisches Denken strukturiert haben und mir über das begrenzte Thema der Arbeit hinaus stets den Horizont offen gehalten haben. Auch weitere Kollegen im IFÖE haben mich mit interessanten Einblicken in ihre Arbeit angenehm abgelenkt (z.B. Annette Kolb mit einem Ausflug in die Wälder bei Zeven) oder mit markigen Sprüchen getröstet, wenn die Arbeit mal wieder kaum zu bewältigen war (z.B. Jürgen Meyerdierks: "Man kann viel messen; bevor man zuviel misst, sollte man erstmal das Fieber vom Prof' messen!").

Meinen Bremer Freunden danke ich für entspannte Spielabende und Tanznächte. Enken Hassold, Jana Petermann und Sabine Schäfer danke ich außerdem für die fruchtbaren Korrekturvorschläge zur Einleitung dieser Arbeit.

Meinen Eltern danke ich für das offene Ohr während unserer zahlreichen langen Telefonate, für die entspannten Wochenenden in Aachen und für die Finanzierung des kleinen gelben Wagens, der mir so viele Geländefahrten erheblich erleichtert hat.

Daniel Nakhla danke ich für die liebevolle, fachlich distanzierte Teilnahme an meiner Arbeit, für die Kraft spendenden Weisheiten von Nietzsche und den Zen-Meistern sowie ein partnerschaftliches Zusammenleben, dass mir die Haushaltsführung gerade während der turbulenten Laborphasen erleichterte und mich vor der Einsamkeit in der Schreibphase bewahrte. Noah danke ich für die heitere Wahrheit seines Kindermunds (*„Musst du eigentlich IMMER nur arbeiten? IMMER nur an dem Computer? Kannst du gar nichts anderes?"*).

Zuletzt danke ich meinen sieben Chefs beim Naturschutzbund (NABU) Rheinland-Pfalz (Werner Kern in Landau-Mörzheim, Franz Grimm in Gleisweiler, Joachim Zürker in Zeiskam, Herbert Magin in Westheim, Hans Frech in Göcklingen, Monika Bub in Haßloch und Siegfried Schuch in Mainz). Sie gewährten mir eine weitgehend freie Zeiteinteilung bei meiner neuen, ausfüllenden Aufgabe in der NABU Regionalstelle Südpfalz. Nur dank des meist späten Arbeitsbeginns und spontaner freier Nachmittage bei Deadlines der Zeitschriften konnte ich die Dissertation relativ zügig zu einem erfolgreichen Ende bringen.

Annexe

Friedrich Nietzsche: Das Können, nicht das Wissen, durch die Wissenschaft geübt

"Der Wert davon, dass man zeitweilig eine strenge Wissenschaft streng betrieben hat,

beruht nicht gerade in deren Ergebnissen:

denn diese werden, im Verhältnis zum Meere des Wissenswerten,

ein verschwindend kleiner Tropfen sein.

Aber es ergibt einen Zuwachs an Energie, an Schlussvermögen, an Zähigkeit der Ausdauer;

man hat gelernt, einen *Zweck zweckmäßig* zu erreichen.

Insofern ist es sehr schätzbar, in Hinsicht auf Alles, was man später treibt,

einmal ein wissenschaftlicher Mensch gewesen zu sein."

aus:Friedrich Nietzsche: Werke I: Menschliches, Allzumenschliches und andere Schriften. Könemann , Köln 1994